The Institute of Biology's
Studies in Biology no. 100

Cellular Recognition Systems in Plants

J. Heslop-Harrison

D.Sc., F.R.S.
Royal Society Research Professor
Welsh Plant Breeding Station, University College of Wales

EX LIBRIS

© J. Heslop-Harrison 1978

First Published 1978 by Edward Arnold (Publishers) Ltd, London
First Published in the USA in 1978 by
University Park Press
233 East Redwood Street
Baltimore, Maryland 21202

Library of Congress Cataloging in Publication Data

Heslop-Harrison, John.
  Cellular recognition systems in plants.

  (The Institute of Biology's studies in biology; no. 100)
  Bibliography: p.
    1. Plant cells and tissues. 2. Cellular recognition. I. Title.
II. Series: Institute of Biology. Studies in biology; no. 100.
QK725.H47   1978   581.8.'76   78-15121

ISBN 0-8391-0250-X

Printed in Great Britain

General Preface to the Series

It is no longer possible for one textbook to cover the whole field of Biology and to remain sufficiently up to date. At the same time teachers and students at school, college or university need to keep abreast of recent trends and know where the most significant developments are taking place.

To meet the need for this progressive approach the Institute of Biology has for some years sponsored this series of booklets dealing with subjects specially selected by a panel of editors. The enthusiastic acceptance of the series by teachers and students at school, college and university shows the usefulness of the books in providing a clear and up-to-date coverage of topics, particularly in areas of research and changing views.

Among features of the series are the attention given to methods, the inclusion of a selected list of books for further reading and, wherever possible, suggestions for practical work.

Readers' comments will be welcomed by the author or the Education Officer of the Institute.

1978

The Institute of Biology,
41 Queen's Gate,
London, SW7 5HU

Preface

The idea that cells and tissues in different parts of the plant or animal body can communicate through the agency of chemical messengers or hormones has been familiar in biology since the early years of the century. In the last two decades ideas about how cells interact at closer range, and indeed when in physical contact, have been taking shape. This text deals with such close-range communication between plant cells.

By selecting examples from various plant groups I have sought to trace some of the common threads, and wherever possible I have attempted to show the connections with parallel work on animal cells. In a rapidly advancing research field like this one speculation tends to outrun fact; but I have not hesitated to refer to some of the theories purporting to account for certain kinds of specific cell interactions. The need for much more work on these interactions in plants to fill some of the many gaps in our knowledge will be obvious enough from the text.

Plas Gogerddan, 1978

J. H. H.

Contents

1 Introduction

1.1 The nature of intercellular recognition systems

In populations of unicellular organisms, cells associate during sexual reproduction and sometimes in such social activities as colony formation. In multicellular plants and animals, contiguous cells of the same individual necessarily interact in the normal course of growth and development; and here once again sexual reproduction requires that cells of different individuals of the same species should come together. Furthermore, cells of *different* species may be brought into close association in such relationships as commensalism, symbiosis and parasitism. In the various situations the interactions may be of the most general kind, or they may be quite specific, occurring only between particular types of cell and leading to responses of a highly characteristic kind in one or both partners. A cell that reacts in a special way in consequence of association with another must do so because it acquires 'information' from that other, information that must be conveyed through chemical or physical signals. In the shorthand of biological parlance, it is customary to refer to phenomena of this kind as 'recognition', in analogy with the way human individuals 'know' each other in the population at large. The terminology is not altogether satisfactory because human beings have to 'get to know' each other before there can be mutual recognition. Interacting cells behave in their characteristic ways because they are already programmed to transmit and receive their special signals. There is no learning stage, so that 'cognition' might be a better term for the form of awareness that one exhibits of the other.

Much the greater part of our present knowledge of the intimate details of cellular interactions has come from work on animal cells, which often reveal what appear to be cognitive properties in very dramatic ways. During embryo formation, cells make coordinated movements in the formation of organs, and in doing so they associate preferentially with other cells and become distributed in precise patterns within the population as a whole. An informative experiment illustrating the powers of sorting out and reassembly can be done with cells taken from embryos into culture. Artificial mixtures of cells from two different kinds of tissue do not form stable mixed aggregates; instead the cell types redistribute themselves in the course of time so that the classes separate. It is indeed as though the cells recognize like and unlike and undertake movements to maximize the degree of association within each class. Just *how* this works is still unclear, but the experiment does highlight one important fact, namely that the properties of the cell surface are very significant in the social behaviour of animal cells. The cells sort themselves out and then adhere in more or less homogeneous aggregates, cell membrane to cell membrane. Adhesiveness evidently plays an important part in the whole complex process, and one can see that the specificity of the associations could itself be governed through modulations of the adhesive property. The mutual 'stickiness' might differ because at a molecular level there are precisely defined complementary binding sites, varying between cell classes. Or, possibly, the differential aggregation could result from different distributions or densities of binding sites without necessarily any great chemical differentiation.

The fact that animal cells do bear identifying markers is now well established from research on the surface antigens. One class of these is controlled by a group of major genes, the histocompatibility complex. The products of this complex participate in the control of certain types of interaction, being concerned, for example, in the discriminations between 'self' and 'non-self' that lead to the rejection of foreign tissue in grafting. A current theory of the evolution of the histocompatibility system holds that it is basically concerned with the control of development, being part of the mechanism responsible for the cell recognitions thought to play a part in embryogeny. Other functions are believed to have been derived from this fundamental one.

There is now a persuasive body of evidence indicating that surface properties also control many of the interactions of plant cells. Yet animal cells and plant cells differ a great deal in the nature of the surface they present to each other and their external environments.

1.2   The cell surface: animals

Earlier interpretations of the structure of the outer membrane of the cell (the plasmalemma), and indeed of the membranes within the cell, were based upon a model proposed in 1935 by J. F. Danielli and H. Davson. According to this model, the lipids (mainly phospholipids) of the membrane are distributed in two layers so that the hydrophobic 'tails' of the molecules are directed inwards towards each other and the hydrophilic 'heads' outwards to the aqueous phase of the cytoplasm on the one side and the intercellular space on the other. The proteins of the membrane are envisaged as being mainly distributed in two layers on the hydrophilic surfaces. This structure meets thermodynamic criteria for stability, and is consistent with various biophysical properties established both for membranes *in situ* and those separated from the cell. Moreover, electron microscopic images of animal and plant cell membranes generally show a three-layered structure readily reconciled with the Danielli–Davson model, a fact that led to the 'unit membrane' concept of J. D. Robertson, advanced in 1959.

The knowledge that biological membranes change their physical and chemical characteristics during the normal life of the cell, seen for example in variation in their permeability properties, has all along made it necessary to assume that the structure is to some extent dynamic. During the 1960s new evidence accumulated from many sources bearing on the nature of animal cell membranes, and this led to a reappraisal of their organization, culminating in the *fluid mosaic* model of SINGER and NICOLSON suggested in 1972. This preserves a basic feature of the Danielli–Davson model, the lipid bilayer, but proposes that the membrane-associated proteins are of two kinds, those embedded in the lipid bilayer or stretching through it (integral membrane proteins) and those distributed on the surface (peripheral membrane proteins). The special feature of this model and the one that required the most far-reaching re-consideration of membrane properties was the postulate that the integral membrane proteins are free to move *laterally* in the lipid layer. Their stability within the thickness of the membrane is presumed to be determined by an asymmetric distribution of hydrophilic and hydrophobic groups in each protein molecule, a distribution which establishes that part is firmly embedded in, or oriented across, the lipid bilayer, with part protruding from it. No forces constrain the protein molecules laterally, however, other than those they develop between themselves, or which may act from either side of the membrane on the protruding parts. A diagram of the fluid mosaic model of the membrane is given in Fig. 1–1.

The fluid mosaic model comfortably accommodates several features of the outer membrane of the animal cell established through recent research. The membrane lipids include glycolipids with short chains of sugar residues in the molecule; these would be disposed with the oligosaccharide portion extending into the aqueous phase and the acyl chains directed into the membrane. Among the most important classes of

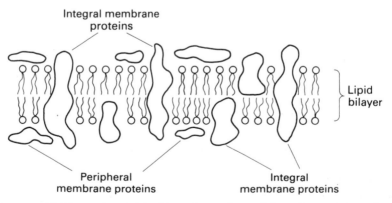

Fig. 1–1 Fluid mosaic model of the cell membrane (plasmalemma). (Based on SINGER and NICOLSON, 1972.)

membrane proteins are the glycoproteins, proteins with simple or branched carbohydrate chains covalently attached. The distribution of these is indicated in Fig. 1–2. The carbohydrate portions of glycolipids

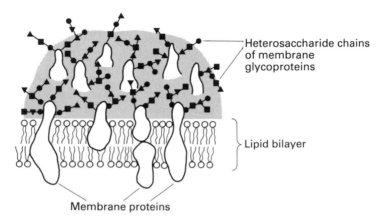

Heterosaccharide chains of membrane glycoproteins

Lipid bilayer

Membrane proteins

Fig. 1–2   Profile and surface of the cell membrane to show the possible disposition of the glycoproteins. The integral membrane glycoproteins emerge from the lipid layer, with the heterosaccharide chains forming the sugar-rich coating.

and glycoproteins form a sugar-rich coating on animal cells, to which the name glycocalyx has been given. With some cell types, the coating can be made visible in both light and electron microscopes by suitable staining procedures. The dye alcian blue stains acid polysaccharides, and cells stained with this often show the coating as a fuzzy layer on the outer face of the cell membrane, extending outwards with diminishing density for as much as 100 nm.

   The lateral mobility of proteins and glycoproteins within the membrane is strikingly demonstrated by the movement of surface markers on living animal cells. Lectins, proteins the properties of which are discussed further in section 1.4, bind preferentially to specific sugar groupings on animal cell surface. Lectin molecules can be labelled with fluorescent dyes, and their distribution after attachment to a cell can then be observed directly by fluorescence microscopy. It is commonly found that, immediately after binding, the fluorescent lectin is randomly distributed over the cell surface, but very soon the labelled molecules begin to clump together and often they all end up as a 'cap' over one pole of the cell. This effect is best explained on the assumption that the receptors for the lectin held in the membrane, presumably the integral glycoproteins shown in Fig. 1–2, migrate laterally in the lipid bilayer after the binding.

   Animal cell membranes are thus 'dynamic' in the sense that movements of molecules take place freely within them. They are dynamic also in another sense, in that the characteristics of the coating change during development in such a way as to suggest that the exposed surface carbohydrates of the membrane glycoproteins undergo variation during the life of the cell. In fact, there may be a continuous and quite closely controlled flux of these components of the cell surface. The glycoproteins are probably synthesized in the endoplasmic reticulum and then passed through the Golgi apparatus of the cell, the sugars being attached to the protein later in the processing sequence by specific glycosyltransferases. They reach the cell surface with the discharge there of the dictyosome vesicles. These vesicles contribute new membrane to the plasmalemma, which itself is thus in a continuous state of change. This process provides one means through which the surface carbohydrates can be varied according to the metabolic state of the cell. Some observations suggest that changes may also be brought about by other processes after the incorporation of the glycoproteins into the cell membrane through the addition or excision of sugar groups.

1.3   The cell surface: plants

There is no reason to believe that the membranes of plant cells differ in any basic way from those of animal cells: they have broadly similar chemical compositions, and show much the same range of physical properties. However, the plasmalemma of the plant cell abuts a cell wall, and it is this fact that accounts for many of the differences between plant and animal cells. The primary wall, in general, consists of a microfibrillar component within a less well-ordered matrix material. In all higher plants and many algae the microfibrils are composed of cellulose, a β-1,4-linked glucan, the cellulose molecules in the interior of the microfibril being disposed in a crystalline manner with those towards the outside arranged more randomly. The microfibrils, which are 10 nm or more in diameter and of indeterminate length, are held in the matrix, itself composed of hemicellulose and pectic substances. Hemicellulose is the name given to a class of branched heterosaccharides in which the sugars are mainly xylose, mannose, galactose, arabinose and glucose, the combinations varying according to the taxonomic group and also sometimes according to the developmental state of the cell. The pectic substances are gel-forming polysaccharides with galacturonic and glucuronic acids in main and side chains together with arabinose, xylose, rhamnose and other sugars.

In so far as the primary wall of the plant cell is carbohydrate in nature it is in some sense comparable with the carbohydrate coat of the animal cell, and the fact has prompted the extension of the concept of the glycocalyx to cover both. However, for the older plant cell the analogy cannot be pressed too far. The heterosaccharides of the animal cell coat are parts of the membrane glycoproteins and are thus anchored into the membrane. Even though they may be lost slowly into the intercellular spaces by

ablation from the membrane due to the continuous turnover going on at
the cell surface, they can scarcely be said to form a wall in the sense of the
plant cell wall. This is much thicker, and has considerable mechanical
strength so that it forms a box enclosing the protoplast. One aspect of the
different relationship is seen in the fact that the plasmalemma can be
withdrawn from the wall of the mature plant cell by plasmolysis.
Nevertheless, in the young cell the relationship of the cell surface with the
wall may be very close indeed. The precursors of the cell wall polysaccharides are passed into the young wall by the activity of the Golgi
apparatus of the cytoplasm, the dictyosome vesicles once again contri-
buting portions of membrane to the plasmalemma as this happens.
Moreover, the enzymes responsible for the synthesis of the cellulose
microfibrils appear to be clustered in particles or aggregates held on, or
more probably partly in, the cell membrane. Interestingly, the assembly
of the long microfibrils may require the slow, controlled migration of the
enzyme aggregates in the membrane, a concept entirely compatible
with the fluid mosaic model. Also, at least in the younger wall, glycoproteins are often present, and sometimes abundantly so. These must
be transferred through the plasmalemma, and presumably they are
synthesized in much the same manner as the glycoproteins of the animal
cell surface. A generalized diagram of the plant cell wall is given in Fig.
1–3.

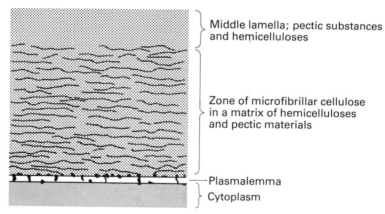

Middle lamella; pectic substances
and hemicelluloses

Zone of microfibrillar cellulose
in a matrix of hemicelluloses
and pectic materials

Plasmalemma

Cytoplasm

Fig. 1–3   Diagram of the primary wall of the plant cell.

The presence of walls necessarily precludes a direct membrane-to-
membrane contact for most classes of plant cell. The exceptions are found
among some lower groups, and universally in at least one phase of sexual
reproduction in all plants – that stage when the gametes, or the cells
carrying the gametes, come together as a prelude to fertilization. Even in
reproduction, however, the initial overtures – comprising the courtship,
so to speak – generally occur between walled cells. So in reproductive as

in somatic interactions plant cells mostly have to negotiate through polysaccharide walls. This may be achieved through the protoplasmic connections formed by the plasmodesmata; these give not simply membrane contact, but membrane continuity between contiguous cells. Or there may be one-way or reciprocal signalling through the agency of soluble, diffusible messengers, originating in one cell, passing across the wall and being received by receptors on or in the other. Or yet again the function of the cell membrane in communication may be transferred spatially so that some of the interactions are on the outer surface of the wall, secondary systems alerting the cell within as to what is going forward. Some of the possibilities in the tissues of a multicellular plant are summarized in a simple form in Fig. 1–4. It can be seen from a model of

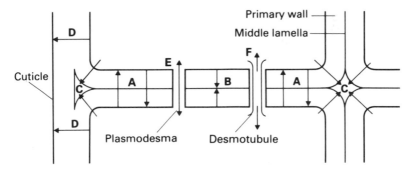

Fig. 1–4 Pathways of communication between plant cells. Signals can be passed by: A – diffusion from plasmalemma to plasmalemma across both intervening walls; B – diffusion from neighbouring cells, with interactions in the wall; C – diffusion into and through intercellular spaces; D – diffusion through the cuticle onto the outer surface; E – movement through plasmodesmata; F – movement through desmotubules.

this kind that close-range cell communication can be viewed as but one element in the wider system where diffusing molecules derived from remote tissues, the mobile hormones, also fulfil their functions.

1.4   A word about lectins

It so happens that one class of molecules produced by plants has played a very significant part in the investigation of cellular interactions in animals, namely the lectins. Lectins, sometimes also known as phytohaemagglutinins, were first detected almost a century ago as compounds, produced by plants, capable of causing the coagulation or agglutination of red blood cells. Today many hundreds of such compounds are known, from a thousand or so species of flowering plants, with the capacity of agglutinating various types of animal cells, and several have been typified chemically and structurally. Among those most commonly used in experiments on animal cells is concanavalin A (con A), from the jack bean, *Canavalia ensiformis*. The structural unit of con A is a protein with a molecular weight of about 25 000 daltons. These units may be associated to form dimers or tetramers or larger aggregates, depending on the pH of the environment. The structural units are asymmetrical in shape, and each has a pocket in the molecule that can accommodate a sugar residue, the dimer thus having two and the tetramer four such binding sites. The binding property of con A is specific towards a-D-mannose and a-D-glucose residues, and the molecule attaches specifically to saccharides containing these residues, including those forming parts of glycoprotein molecules. It is to this property that con A owes its capacity for agglutinating animal cells. Cells with accessible surface saccharides with sugars meeting the specificity criteria will bind con A, and since the dimer is 'bivalent' so far as its binding capacity is concerned, the molecule can link the first cell to another with the same surface sugar groups by attachment at the second site. In suitable concentrations, con A can cause the aggregation of a whole population of cells, the agglutination response. The specificity can be checked because the binding cannot take place when the appropriate sugars are present in quantities sufficient to preoccupy the binding sites, in the case of con A, that is, mannose or glucose.

Other plant lectins, some themselves glycoproteins, show different binding specificities, and some are known that seem to recognize and attach to combinations of sugars terminating, or intercalated in, heterosaccharide chains. The binding affinity is reminiscent of that of enzyme for substrate, although the attachment is more persistent. The union is not due to the formation of covalent bonds, however, and it can be broken by appropriate treatments.

Because of their special properties, lectins provide a valuable method of 'probing' the cell surface, since their attachment can be taken to indicate the presence of the sugars for which they have specificity. In some instances the association with a lectin provokes far-reaching changes in the behaviour of the cell. Con A, for example, induces lymphocytes to undergo mitosis, and when applied to certain types of egg cell, can inhibit fertilization.

Like other proteins, lectins are probably synthesized in the endoplasmic reticulum of the plant cells. It is known that some at least are transferred into the cell wall, where they accumulate sometimes in surprisingly large amounts. Later some of the implications of lectin effects will be examined and the possible functions of these molecules in plants considered.

1.5 Types of recognition response in plants

The classical literature of botany is replete with examples of interactions which, with the wisdom acquired from recent research, one can readily see must depend on recognition responses of greater or lesser specificity. Events associated with reproduction have already been mentioned. In every plant group with a sexual process there must be some selectivity at certain stages in the association of the conjugating partners. The gametic fusion of unicells clearly has to be selective, and the same must be true for the fusion of free-swimming gametes produced by multicellular algae and fungi like *Ulothrix* and *Allomyces*. The fertilization of static eggs by motile spermatozoa as in plants as widely diverse as *Vaucheria*, *Fucus*, *Monoblepharis*, *Funaria*, *Pteridium* and *Cycas* clearly requires a combination of chemotactic guidance together with more specific controls determining the kinds of cellular unions to be permitted. The same must apply when gamete nuclei are brought together through the association of non-motile cells, as in the contact of spermatium and trichogyne in the red algae, or in the conjugation of different filaments in *Mucor* and *Spirogyra*, where short-range interactions induce special patterns of growth in contiguous cells. In the higher gymnosperms and in the angiosperms the capture of pollen and the subsequent choice between acceptable and non-acceptable pollen tubes are phenomena of the same general kind.

Powers of discrimination may be seen in many other types of association. The choice of partners in the lichen symbiosis must depend on mutual recognition; and higher in the evolutionary scale of plants we can suspect the same in the association of host-specific dodders and their victims. The relationships of vascular plants and viral, bacterial and fungal pathogens embrace another whole field of recognition phenomena, where host specificity, resistance and susceptibility all imply specific interactions. In less natural situations, the results of artificial grafting where the genetic constitution of stock and scion determines whether the union will take or not suggest the operation of some form of cellular recognition system.

Grouping all of these different classes of phenomena under the general heading of recognition events acknowledges the feature they share in common, namely the expression of specificity or selectivity. This certainly cannot be taken to imply that the underlying controlling systems are identical or necessarily even broadly similar – even, say, in the interactions between gametes in the different groups. Nevertheless, a body of evidence is now accumulating which suggests that many responses do have much the same kind of molecular basis, and it is an intriguing fact that this may not be greatly different from that believed to underlie many recognition events in animal cells and tissues.

2 Recognition Systems in Algae and Fungi

2.1 Introduction

Algae and fungi have many advantages for the investigation of short-range cellular interactions. Their body-forms are simpler than those of the higher plant groups so that processes like fertilization can be more easily observed; many can be taken through complete developmental cycles under controlled laboratory conditions; and, where mass cultures can be raised, adequately large samples can be acquired for chemical analysis. It is therefore not surprising that a good deal of our present knowledge of cellular recognition systems in plants has come from these lower groups. There remains the question of how much use the gained knowledge is likely to be in seeking to understand the more complicated types of cell interaction found in the more advanced groups, particularly in the angiosperms. It is probable that the evolution of more elaborate body forms and more complex tissue systems has been accompanied by – indeed, has even been dependent upon – the development of a more sophisticated mechanism for cell communication. But one might at least expect the systems found among lower plants to be among those available to plants higher in the evolutionary scale, and it is therefore attractive to look for parallels.

2.2 Algae

2.2.1 Gamete fusion in Chlamydomonas

Chlamydomonas, a genus of unicellular freshwater alga belonging to the Chlorophyceae, has a sexual process in which morphologically similar, biflagellate gametes fuse to give the zygote. The gametes are transformed vegetative cells, formed when a population experiences nutrient deficiency. The gametes are not differentiated structurally, but many species show heterothallism, the gametes belonging to different mating types, designated as (+) and (−). No conjugation takes place within a gamete population of a single mating type but when (+) and (−) populations of gametes from the same species are mixed, agglutination begins. Clusters of gametes are formed containing mixtures of the two mating types, the clumping being due to adhesion between the tips of the flagella. Some time after this event, the clusters break up into pairs of gametes, each with a (+) and a (−) partner. The cells then link by the body region; the cell walls fuse, a conjugation tube is formed, and fusion of the cytoplasms and nuclei follows. Several features of this process make it a favourable one for the study of contact and recognition events. Vegetative cells do not agglutinate, and because the transformation into gametes can be controlled, the development of the recognition system can be followed. The surfaces that come into contact during agglutination are precisely known: they are the membranes of the flagella tips. Finally, the responses are very specific: the fusion is between gametes of different sex, and beyond this there is species specificity, gametes of different species, with certain exceptions, being unable to fuse with each other whatever the mating type. These relationships are shown for three species in Table 1.

Table 1   Sex and species specificity in the mating of *Chlamydomonas*. A = agglutinuation when the cultures are mixed; O = no agglutination when mixed. (Data from WIESE, 1974.)

Species	*C. eugametos*		*C. moewusii*		*C. reinhardi*	
	(+)	(−)	(+)	(−)	(+)	(−)
C. eugametos						
(+)	O	A	O	O	O	O
(−)	A	O	O	O	O	O
C. moewusii						
(+)	O	O	O	A	O	O
(−)	O	O	A	O	O	O
C. reinhardi						
(+)	O	O	O	O	O	A
(−)	O	O	O	O	A	O

Most of our knowledge of the details of the conjugation process in *Chlamydomonas* has come from the work of Wiese and his collaborators (e.g. WEISE, 1974). One informative finding was that cells in the gametic state shed into the medium material which itself is capable of causing agglutination of gametes of opposite mating type; the material from (+) cells agglutinates (−) cells and the material from (−) cells agglutinates (+) cells. Agglutination of (+) gametes of *C. eugametos* by the material from the culture medium of (−) gametes is illustrated in Fig. 2–1. After this homosexual agglutination there is, of course, no cell fusion. These materials, referred to as isoagglutinins, have been shown to contain glycoproteins. They carry with them not only the *sex* specificity, but also the *species* specificity (Table 1). Their involvement in the adhesion and recognition responses is not therefore in doubt.

The analogy with animal-cell agglutination immediately suggests the possibility that the adhesion of the gametes might be something to do with membrane-held glycoproteins. Evidence favouring this idea comes from the fact that the lectin, con A, affects the agglutinability of the gametes. In high concentrations it brings about isoagglutination of

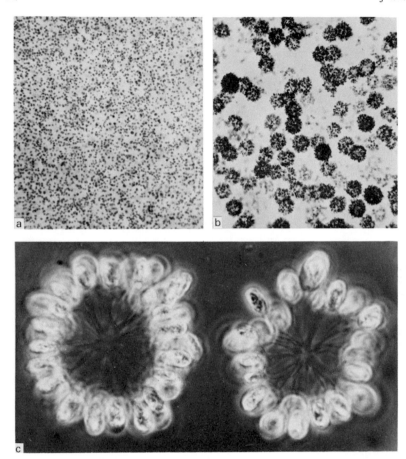

Fig. 2–1 Agglutination of (+) gametes of *Chlamydomonas eugametos* by isoagglutinin from the culture medium of (−) gametes. (a) Free-swarming cells in the absence of the isoagglutinin; (b) agglutination a few minutes after the addition of the isoagglutinin to the medium; (c) detail, showing two clusters of the agglutinated gametes as seen with dark-field illumination. The cells are held together in the rosette by the adhesion of the tips of the flagella. The body of each cell is about 12 μm long. (Micrographs reproduced by the courtesy of Professor L. Wiese.)

gametes of both sexes of certain species, and in the same species at lower concentrations it destroys the capacity of one sex to bring about agglutination when mixed with the other (Table 2). Con A does not induce isoagglutination when mannose is present, indicating that the effect depends on the specific sugar-binding capacity of the lectin. Furthermore, pre-treatment with α-mannosidase, an enzyme that might be expected to remove α-D-mannose residues from the flagellar surface,

Table 2   Effect of the lectin, concanavalin A, and certain enzymes on the agglu-tinability of (+) and (−) gametes types of three species of *Chlamydomonas*. S= agglutinability lost after treatment when tested against untreated gametes of other mating type; O = agglutinability not lost. (Data from WIESE, 1974.)

Treatment	Species and mating type					
	C. eugametos		*C. moewusii*		*C. reinhardi*	
	(+)	(−)	(+)	(−)	(+)	(−)
Concanavalin A	S	O	S	O	O	O
Trypsin	O	S	O	S	S	S
α-mannosidase	S	O	S	O	O	O

also affects the agglutinating power of just those gametes whose agglutinating capacity is also affected by con A (Table 2). These results suggest that mannose is present in the combining sites of both classes of gametes from *C. eugametos* and *C. moewusii*, but that this is important for conjugation only on the (+) mating types. This would be expected were the complementary sites concerned in the sexual contact on the (−) gametes proteins with binding sites for mannose, like con A itself. This conclusion is supported by the finding that proteolytic enzymes like trypsin destroy the agglutinating power of the (−) gametes (Table 2). For *C. eugametos* and *C. moewusii*, Wiese has concluded that sex cell contact involves at some point a protein at the combining sites of the (−) gametes and a carbohydrate with mannose in a terminal position at those of the (+) gametes (Fig. 2–2).

What then of the third species of Table 1, *C. reinhardi*? In this species neither gamete type is inactivated by con A in low concentrations, although the gametes of both types are agglutinated by high concentrations. On the other hand, both gamete types lose their agglutinating powers when exposed to trypsin. Evidently a different recognition code operates here, although it is not exluded that adhesion between gametes depends upon glycoproteins with different surface sugar residues, in one or other partner, for which con A has no affinity.

The cell-free isoagglutinins from the culture media of *Chamydomonas* gametes have been investigated both chemically and structurally. The glycoproteins from those of different species differ both qualitatively and quantitatively in the sugars present in the molecules. These differences could be connected with the species-specificity of the mating reactions, but there is as yet no evidence for this. However, it is now known that the isoagglutinins, which have been said to have very high molecular weights in the region of 1×10^8 daltons, are themselves not molecules but pieces of membrane. Viewed with the electron microscope after separation by density gradient centrifugation they are seen to be vesicles with fuzzy coats, and the coats may well be the sites of the factors concerned with recognition and adhesion. These vesicles are derived from the

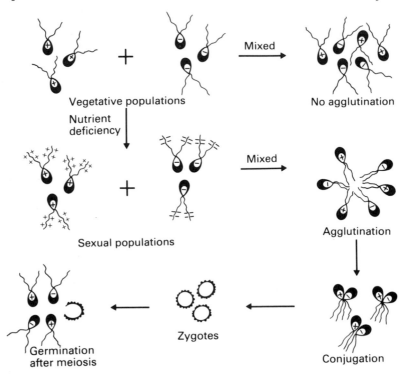

Fig. 2–2 Gamete formation, agglutination and conjugation in *Chlamydomonas*.

membranes of the flagella, and it seems likely that they are shed continuously at the tip, the membrane being replenished by the flow of new membrane substance from the base. A mechanism of this kind would account for the way in which the cell is able to change the properties of the flagellar membrane during development, as in the transformation of a vegetative cell into a gamete. The factors concerned in recognition and gamete adhesion are presumably held on the outer surface of the membrane, and the configuration of the surface proteins and the nature and distribution of the carbohydrate portions of glycoproteins presented to the environment at any one time would depend on the metabolic activities of the cell during the synthesis of the particular bit of membrane currently exposed, itself determined by which genes happened to be available for transcription.

Chlamydomonas, then, evidently has a sophisticated signalling and recognition system working through the properties of the cell surface. Much still remains to be found out about how it operates, and especially how it provides for discrimination between different species, but the fact that it exists immediately alerts one to the possibilities open for the operation of similar systems in higher green plants.

2.2.2 Fertilization in Fucus

While *Chlamydomonas* is isogamous, brown algae of the Fucales are oogamous, but here also the gametes fuse when free in the bathing medium, in this case sea-water. The large, non-motile eggs are fertilized by free-swimming sperms, and very soon afterwards the zygote begins to form a polysaccharide wall. Work by BOLWELL *et al.* (1978) has shown that fertilization is strongly species specific. In experimental conditions, freshly released eggs of *Fucus vesiculosus*, a common coastal seaweed, are never fertilized by sperms of *F. serratus* even when every opportunity is given, nor is the reciprocal fertilization possible. Similarly, neither species can be fertilized by a species of a neighbouring genus, *Ascophyllum nodosum*. Experiments with *F. serratus* showed that fertilization is controlled by determinants on the surface of the gametes. Con A and another fucose-binding lectin, bind specifically to the eggs at low concentrations, and the binding inhibits fertilization (Fig. 2–3). At higher concentrations these lectins are also attached to the sperms, again inhibiting fertilization. The lectin-binding sites are not distributed over the whole of the egg surface, but are concentrated in one small area. The block to fertilization caused by the lectins appears to result from competition for binding sites in this zone. The experimental findings are consistent with the view that the attachment of the sperms in the

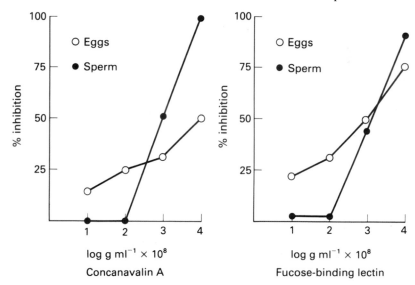

Fig. 2–3 Effect of two lectins, concanavalin A and fucose-binding lectin, on fertilization in *Fucus serratus*. When the eggs were exposed to the lectins before admission of the sperms, fertilization was inhibited to some extent even at the lowest concentrations tested. Sperms were inhibited at higher concentrations. (Data from BOLWELL *et al.*, 1978.)

fertilization site depends on the presence of surface carbohydrates, and one might deduce from the lectin specificities that these carry exposed fucose residues and also either mannose or glucose. In agreement with this interpretation is the observation that α-fucosidase, an enzyme that might be expected to remove α-L-fucose groups from the saccharide portions of membrane glycoproteins, also eliminates the capacity of eggs to accept sperms.

The parallel with *Chlamydomonas* is clear enough. In each case the adhesion of the gametes depends on the properties of particular parts of the cell membrane, and these properties appear to be determined by surface carbohydrates on the one partner and, complementing them, carbohydrate-binding proteins on the other.

In *Fucus*, as in *Chlamydomonas*, the basis for species specificity remains to be worked out, and it cannot yet be said whether, in this recognition system, selectivity depends on the sugar sequences of the surface glycoproteins, or whether some quite different properties of the membranes are involved.

2.2.3   *Pheromones and inducers in algal reproduction*

In the two examples just considered, the recognition of species and mating type takes place when the cells come into physical contact. In *Chlamydomonas* this is probably a chance event, but in *Fucus* the sperms locate the egg chemotactically, swimming towards the source of an attractant diffusing from it. The egg exudates are of course sex specific, but at least within genera probably not markedly species specific. They act as pheromones, providing a means of communication between egg and sperm, much as air-borne insect pheromones function in permitting the males to locate the females. Agents that act in a similar way as chemical signals but with still more far-reaching effects are known in other algae. *Volvox* (Chlorophyceae) is a colonial form, the coenobium or colony being composed of hundreds or thousands of cells arranged in a hollow sphere. Like *Fucus*, the genus is oogamous. The colonies are male and female, or hermaphrodite, according to species. The remarkable feature in our present context is that female colonies can induce vegetative colonies of the same species to pass into a reproductive state and to act as males, producing packets of sperm. In *Volvox aureus*, the agent responsible produced by the female colonies (male inducing substance, MIS) has been partly characterized and shown to be a protein. When the sperms are released, they are attracted chemotactically to the eggs of the females, demonstrating yet another level of chemical signalling.

The mutual induction of major morphogenetic processes is to be seen in the filamentous green algae of the Zygnemales, although nothing is known yet about the inducing agents. In this group conjugation takes place between neighbouring cells of the same or different filaments with the formation of conjugation tubes through which the protoplasts pass in an amoeboid manner to fuse and give the zygotes. *Spirogyra*, a familiar genus of the order, commonly shows so-called 'scalariform' conjugation, in which the cells of laterally apposed filaments of different mating types pair two by two. The process is precisely controlled. Neighbouring cells produce accurately aligned papillae, and these grow out to make tip contact. As in the pairing of *Chlamydomonas* cells, the wall in the contact area is then digested away to give the cellular continuity. Should there be a misalignment of the cells in the apposed filaments, two cells of one may attempt to conjugate with one of the other, but once the first fusion is achieved, the second outgrowth is inhibited. The conjugating filaments are ensheathed in a mucilaginous matrix, which no doubt helps to ensure that they are held within close enough distance to permit the interaction of the cells. This interaction is presumably mediated by pheromones exchanged through the matrix, and the fact that these are effective only between closely apposed pairs of filaments suggests that they have very short diffusion ranges or are effective only in quite high concentrations, attained only in the immediate vicinity of the cells from which they are released. Even when closely apposed, cells of different *Spirogyra* species in mixed colonies rarely interact to form conjugation tubes. The recognition response between the filaments therefore shows both mating-type and species specificity.

The various algal examples illustrate an important principle, namely that the cellular interactions involved in reproduction are often manifold, and can operate at different levels of specificity. This is seen also among the fungi.

2.3   Fungi

2.3.1   *Mating behaviour in yeasts*

Many of the cellular interactions associated with reproduction in the algae are matched quite closely in the fungi. The unicellular yeasts (members of the Ascomycetes), for example, provide an intriguing counterpart to *Chlamydomonas*, exhibiting at the same time several unique features. *Saccharomyces cerevisiae*, brewers' yeast, has homothallic and heterothallic strains, and the sexual interactions of the latter have been studied in considerable detail.

As in *Chlamydomonas*, cells of the two mating types in the appropriate developmental state agglutinate when mixed in suspension cultures, the contacts here being anywhere over the cell surface. After the aggregation, tubes are developed from the walls of adjacent cells of different mating type, continuity between the cytoplasms is established and the nuclei then fuse. The preliminaries to the conjugation are very complex. In an active population, the cells will be undergoing division asynchronously. They will therefore be at all stages of the cell cycle, some with a DNA content corresponding to a single chromosome set (G_1), some in the course of DNA synthesis (S) and some with replicated DNA (G_2) just before passage into division (M). An early effect of mixing different mating types is to induce synchrony in the potential partners, which stabilize in the G1 state (HARTWELL, 1973). Furthermore, if one or both of the prospective partners are not initially in the agglutinating state, this is attained after a short period in mixed culture. Agglutination and cell fusion then follow.

DUNTZE, MACKAY and MANNEY (1970) showed that the steps preparatory to agglutination depend on diffusible sex factors (pheromones) produced by the interacting cells. The same inducing factors are probably concerned in both parts of the response: cell synchronization and the attainment of agglutinating capacity. Yanagishima and colleagues (e.g. YANAGISHIMA *et al.*, 1976) have indicated that they are peptides, although there has been no complete chemical characterization yet. In any event, each is highly specific in its effects, acting only on the mating type opposite to that from which it was released. In a mixed population the mutual interaction creates the conditions for cell fusion.

The agglutination response has been studied most fully in another yeast, *Hansenula wingei*. In this species specific diffusible sex factors like those responsible for inducing the agglutinating state in *Saccharomyces cerevisiae* have not been found. However, when haploid cells of the two mating types are mixed in an appropriate medium, agglutination does not begin immediately, but only after a lag period of an hour or so. This suggests that some sort of mutual induction does occur, and it has been found by BROCK (1961) that if inhibitors of protein synthesis are introduced during the lag period agglutination is prevented, suggesting that the production of new proteins is essential for the reaction. Study of the agglutination process (reviewed by CRANDALL and BROCK, 1968) has shown that the adhesion of the two mating types is brought about by the interaction of complementary glycoproteins present on the cell surfaces. Since these are distributed over the whole of the wall one cell can adhere to several others of the opposite mating type. The cell is thus 'multivalent' in the same manner as the con A molecule with its two or four sugar binding sites, but of course on a much larger scale. Because of this capacity large numbers of cells aggregate in a mixed culture, giving the agglutination response.

The glycoproteins on the surfaces of the cells of the two mating types are sex specific. In actively growing cultures in the agglutinating state each can be isolated from the walls and also from the cytoplasm of the cells. Furthermore, the factors are released into culture media, suggesting that they diffuse away from the wall and are continuously replenished from within the cell. The properties of the factors from the two mating types are different. That from one is itself an agglutinin; it has been isolated by various methods, including affinity chromatography using cells of the other as the adsorbent. It is heterogeneous in molecular size, indicating that aggregates of different dimensions are present much as in preparations of the agglutinating factors from the culture media of *Chlamydomonas* gametes. Related to the heterogeneity, no doubt, is the fact that the protein-carbohydrate ratio varies in different preparations. The recognition factor from cells of the other mating type is more homogeneous, and it does not itself act as an agglutinin. Its function in recognition is proved by various properties: it is adsorbed specifically on to cells of the opposite mating type, it neutralizes the activity of the agglutinating factor from the opposite mating type, and when it is stripped from the parent cells these cannot any longer be agglutinated by this factor. The precise nature of the chemical binding, and whether the specificity is determined by a carbohydrate-protein or protein-protein association, remains to be established.

In the yeasts the mating-type substances responsible for agglutination are held on the cell walls. Electron microscopy shows that the cells of each mating type bear a fuzzy external coat, and it is these coats that first bind together during agglutination, suggesting that they are the site of the surface glycoproteins. After the first adhesion, the cells produce outgrowths in the areas of contact. The material of the two walls fuses to form a neck, the partition wall is eroded, and the cytoplasms become continuous. In *Hansenula wingei* none of these growth responses occurs unless cell contact has first been established by agglutination, although agglutination does not necessarily lead to conjugation. In brewers' yeast growing on solid media, cells of complementary mating type that are not actually in contact produce outgrowths directed towards each other. These are presumably chemotropic responses, and their occurrence suggests that the cells produce sex-specific attractants, much in the manner of the conjugating cells of *Spirogyra*. The nature of these pheromones is unknown as yet.

It will be seen from the foregoing that the yeasts have quite complex intercellular signalling systems. Figure 2–4, a composite based on results derived from several species, shows the steps that may be involved, from the initial preparatory phase when the cells of the two mating types are brought into synchrony with the nuclei in the G1 state, to the formation of the conjugation tube and nuclear fusion. Sex specificity is shown during at least three periods.

Under normal growth conditions, the diploid cells produced from the fusion of haploid yeast cells of complementary mating types do not express any of the sex-specific characteristics. When, in due course, the diploid cells undergo meiosis, there is a segregation for sex type in the ascus. The 'packages' of sex-specific characteristics segregate as alleles, but the locus is a compound one, apparently consisting of several tightly linked functional units or cistrons. The independence of some of the functions is shown by the existence of mutants lacking some of the specificities. Mutants are known, for example, which possess the agglutinative capacity but lack the ability to induce the formation of conjugation tubes. This illustrates in another way the multiplicity of the physiological controls involved in the processes summarized in Fig. 2–4.

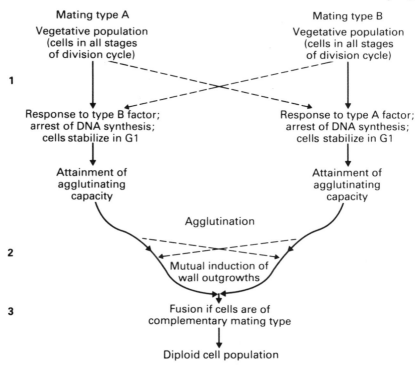

Fig. 2–4 Sequence of events in the mating of haploid yeasts. The first steps, the arrest of DNA synthesis and the reciprocal induction of agglutinability, are known from *Saccharomyces cerevisiae*, and the subsequent ones from *Hansenula wingei*; the diagram is thus a composite. Sex specificity is expressed at three points, 1, 2 and 3.

2.3.2 Sex pheromones in the Mucorales

In the Mucorales the vegetative body is composed of multinucleate hyphae. Sexual reproduction takes place through the development of outgrowths on apposed hyphae, the progametangia. These grow towards each other and the terminal sections are cut off as gametangia; the gametangia then fuse with the confluence of the contents, the fusion product later forming the zygospore. Notwithstanding the differences in cellular organization, there are points of similarity with the conjugation process in algae of the Zygnemales, and again the induction of new, directed growth patterns as a preliminary to sexual fusion suggests control by diffusible pheromones. In heterothallic strains of *Mucor*, and other genera with the same kind of sexual process, progametangia are not formed in cultures of (+) and (−) strains grown in isolation. However, the addition of filtrates of the culture medium of the opposite mating type does induce their formation. At first it appeared that two specific sex substances must be involved in controlling the behaviour, but Van den Ende (reviewed in STEGWEE, 1976) showed that a single compound, trisporic acid, present in the medium of mated cultures, could induce progametangia in both (+) and (−) strains. The sex-specific control turned out be a subtle one. The (+) strain produces a precursor which is converted to trisporic acid *only* by the (−) strain, and reciprocally the (−) strain produces another precursor converted only by the (+) strain. In this instance the sex-specific pheromones, the agents that actually pass between the interacting cells, are not themselves the active factors, but two different precursors of the one active factor.

The chemotropism of the progametangia, when formed, appears to be due not to trisporic acid, but to the diffusion of other pheromones, which may, however, be related chemically. There is still another sex-specific step, since the mere induction of progametangia does not guarantee fusion and the formation of zygospores. These events only happen when the progametangia of opposite mating type come together. Evidently there is recognition on contact, perhaps mediated by materials held in the walls at the tip zone, which is rich in glycoproteins.

The system in the Mucorales is summarized in Fig. 2–5. It will be seen that it matches that of the yeasts in complexity.

2.3.3 *Cytoplasmic recognition systems*

In all of the cases discussed so far the communication has been between cells or walled coenocytes. The effects at different levels of specificity have arisen from the transfer of chemical stimuli from cell to cell or from interactions between cells in physical contact. In the fungi there are some specific recognition phenomena in which the control is not in the wall or cell membrane but is dependent on cytoplasmic contacts. These phenomena appear to play an important part in the reproduction of the most advanced group of the fungi, the Basidiomycetes. In this group there are no specialized sex organs. Hyphae with one nucleus per cell (monokaryotic) fuse to produce a mycelium with two nuclei per cell (dikaryotic). This condition is preserved during further vegetative growth until just before the meiotic division in the basidium, when the nuclei fuse. Plasmogamy, the confluence of the cytoplasms, is not therefore followed immediately by karyogamy. In the genus *Coprinus* (ink caps), as in many others, the fusion of hyphae is rather a random process, and can occur between different species. This indicates that so far as contact and fusion are concerned there is no interspecific incompatibility system. However, when certain pairs of species are tested against each other, it is found that interspecific fusions lead to the death of the hybrid cytoplasm, either immediately or after a few hours. It seems, therefore, that the integrity of species in *Coprinus*, and probably other Basidiomycetes, is guarded by a cytoplasmic recognition and rejection system and not by mechanisms involving the plasmalemma and the wall.

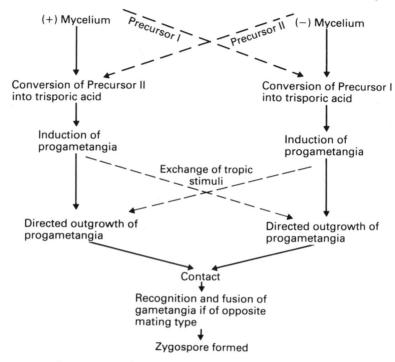

Fig. 2–5 Sex pheromones and recognition in Mucorales.

2.4 Somatic cell interactions in algae and fungi

Many more instances could be cited from algae and fungi of sexual interactions where there is an overwhelmingly strong presumption that the behaviour of partners is governed by intercellular communication of a very precise kind. Fragmentary although our knowledge still is about many of the details, we may yet marvel at the sophistication of some of the systems that have been partly elucidated. There is provision for long-range chemical signalling and surveillance; shorter range tropic or tactic movements leading to docking; reciprocal identification of sex and species; and then the controlled dissolution of separating membranes and walls preparatory to the confluence of cytoplasms and the fusion of gamete nuclei. Interactions of similar kinds in somatic tissues are much more difficult to identify and investigate, and yet it is entirely logical to expect that they could be involved in vegetative development and morphogenesis just as in reproductive events. In this text we are borrowing the current usage of animal cell biology and embryology and using the idea of cellular recognition in the wider sense to cover specific, short-range intercellular communication in general. It is interesting to note that one group of plants – or near-plants – the cellular slime moulds, do indeed exhibit activities quite similar to those seen in animal embryogenesis, and they provide an informative link with other plants of a more orthodox character.

In the vegetative phase the cellular slime moulds form populations of naked, amoeboid cells or myxamoebae. With the transition to reproduction, the myxamoebae aggregate to form a plasmodium, which in due course forms a fructification. The special features of this behaviour are the transition from a population of free-moving cells to what is in effect a single organism, and then the coordinated action of those cells in the often quite elaborate morphogenetic activity of forming the fructification. Two elements of the cellular communication system involved in these processes in certain species have now been identified. The initial 'aggregation' signal is the nucleotide, cyclic adenosine monophosphate (cyclic-AMP). This is released from the first aggregation centres in a rhythmical way, and the signals so generated are passed on and amplified by other cells in the swarm. The signals are picked up by cyclic-AMP receptors in the cell membrane. This form of communication is relatively non-specific, but the later cellular contacts and movements during the morphogenetic stage depend on special properties acquired by the cell surfaces, and these are species-specific. The adhesion of the participating cells appears to follow the presentation at the cell surface of lectin-like molecules, capable of binding with complementary saccharide groups forming parts of glycoproteins on the surfaces of the partner cells. The lectins vary according to species, and their special function in morphogenesis is shown by the fact that development is arrested if the cells are exposed to antibodies specific for their particular lectins.

The fruiting bodies of the larger fungi and the thalli of the larger algae can often be seen to arise through the aggregation of filaments or hyphae, forming the plectenchyma of plant anatomists. The filaments behave in a closely coordinated way during growth and morphogenesis, and a link may be seen here with the way the fruiting bodies of a slime mould arise, although the coordination is in patterns of growth rather than cell movement. In some way accurate 'positional information' is conveyed between the participating filaments. In certain fungi the physical characteristics of the substratum have been shown to affect the growth pattern of hyphae, and presumably such haptotropic movements occasioned by hyphal contacts play some part in the morphogenesis of the more massive types of fungal structure like rhizomorphs and fruiting bodies. But there must also be more subtle forms of communication, and this has been demonstrated in experiments where the growth of hyphae has been shown to be affected by the presence of others separated from them by membrane permeable to small molecules. This is good evidence for the action of pheromones. What part macromolecules may play in the transfer of still more detailed information remains to be discovered.

3 Recognition Systems and Reproduction in Flowering Plants

3.1 The pollen-stigma interaction

The angiosperm life cycle is based on an alternation of generations, the diploid spore-producing plants, the sporophytes, alternating with the haploid gamete-producing plants, the gametophytes. The gametophytes are physically small. The female is the embryo sac, usually with eight nuclei, and the male is the pollen grain, which at the time of dispersal has either two or three nuclei. The pollen grains are produced in the anther following the meiotic division. After conveyance by one or other of several possible agencies to the female flower they are received by the receptive surface of the stigma, an organ of the sporophyte specialized for this function. There each grain germinates to give a pollen tube, an outgrowth of the single vegetative cell of the male gametophyte. The tube penetrates the surface and grows through the style, either in a specialized transmitting tissue or along a canal, until it reaches the ovary. Two male gametes are then released; one moves to the egg and fertilizes it, and the other makes its way to the fused polar nuclei in the centre of the embryo sac where a further nuclear fusion gives the triploid primary endosperm nucleus. The evolution of the pollen tube as a channel for fertilization (siphonogamy) was an event of great significance, since it freed the angiosperms from dependence on water as a medium for fertilization and so contributed to their success as colonists of the land surface. But the adoption of this mode of fertilization also introduced a new kind of physiological relationship. Before the gametes can be delivered, the tube, part of the haploid male gametophyte, must grow like a parasite through the tissues of the diploid sporophyte, and this establishes a kind of interaction not found in lower plants. The cell contacts involved in this phase of the angiosperm life cycle are summarized in Table 3.

Whereas in the algae and fungi considered in Chapter 2 fertilization is regulated by interactions between the gametes themselves or by the haploid tissues that produce them, it will be seen from Table 3 that in the flowering plants this can be achieved by controlling the emergence and growth of the pollen tube as well as by the interaction of the gametes at the actual time of fertilization. In fact, control of the breeding system in angiosperms is indeed most often imposed before the gametes themselves come into close enough range of each other to interact directly; the stigma and style acting as filters to screen out 'unwanted' pollen and pollen tubes before the male gametes are released.

Table 3   Cellular interactions in angiosperm reproduction. (From HESLOP-HARRISON, 1975b.)

Fate of the male gametophyte	Interaction with female-acting sporophyte involves:
Capture	Stigma surface materials
Hydration ⎫ Germination ⎭	Stigma surface and underlying cells
Tube penetration	Stigma cuticle and underlying cells
Tube growth in the style	Transmitting tissue or canal of the style
Entry into the female gameto- phyte and gamete discharge	None: interactions with the female gametophyte

Siphonogamy necessarily involves a whole group of processes in which the male gametophyte and the tissues of the stigma and style must cooperate for success. The pollen must become attached to the stigma, and find there favourable conditions for germination; for example, there has to be a proper osmotic relationship so that the grains hydrate in a controlled manner without either bursting or becoming plasmolyzed. The emerging tube must be able to penetrate the stigma surface, breaking through the cuticle if one is present. After penetration, the tube must be guided through the transmitting tract or canal, and in its further journey be able to take up nutrients after its own reserves are exhausted. At the end of the passage through the style, it must be guided into the micropyle of the ovule and then into the embryo sac. Disharmony between the male gametophyte and the female tissues in any of these processes will lead to failure. *Within* a species natural selection will necessarily work rigorously to preserve the co-adaptation, since without it sterility would follow. But one can see that a sufficiently severe break-down at any point in the co-adaptation *between* species could create a breeding barrier, and establish interspecific incompatibility. This kind of incompatibility often does contribute to the genetic isolation of species. It depends on mal-adjustments between possible partners rather than upon sophisticated recognition mechanisms, but this does not mean that it is any the less effective.

In contrast, closely related species which do not differ greatly in the general physiology of pollen tube, stigma and style often have a type of control where incompatibility seems not to depend simply on maladjustment but upon specific recognition and rejection. Moreover, within species the commonest method of promoting outbreeding and preventing self-fertilization in the angiosperms is through the operation of self-incompatibility systems which discriminate against self-pollen in favour of that from other plants of the same species in the vicinity.

3.2   Self-incompatibility

3.2.1   Genetical control

In the commonest form of self-incompatibility system, the response on the male side is governed by the genotype of the individual haploid pollen grain, and on the female side by the diploid tissues of the style. In the most fully investigated of these so-called *gametophytic* systems, the control is by one gene locus, the *S* locus. There are many *S*-alleles, and the rule is that a pollen tube carrying a given allele is inhibited in a style the cells of which carry the same allele. When there is no match, the pollen tube can grow and effect a fertilization. Thus in a cross-pollination where there was no match between the diploid parents for either *S*-allele all pollen grains would be functional; when both alleles were matched, no pollen would function; and when one allele was matched, half of the pollen grains would be potentially capable of effecting a fertilization and the other half would not. A system of this kind were it to work perfectly would necessarily produce plants heterozygous for the *S* locus in every generation. The scheme is shown in Fig. 3–1. In some families with

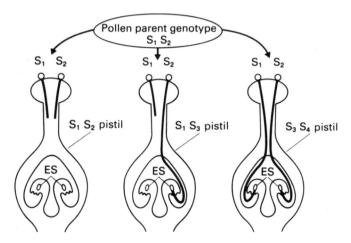

Fig. 3–1   Gametophytic self-incompatibility. A pollen parent of genotype $S_1 S_2$ will be infertile, semi-fertile or fully fertile according to the genotype of the female plant. In most species with the gametophytic system, incompatible pollen tubes are inhibited in the style. ES – embryo sac.

gametophytic self-incompatibility systems, the control is by two or more loci, rejection taking place when each of the alleles in the pollen grain finds a match in the style.

In the second type of self-incompatibility system, the behaviour of the pollen is determined not by its own genotype, but by that of its parent. Again the response on the female side is controlled by the diploid tissues of stigma and style. In this *sporophytic* system there is once again a single S locus, rejection occurring when an allele in the pollen parent is matched in the female tissues (Fig. 3–2).

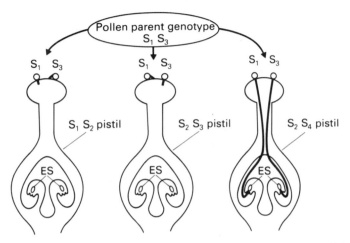

Fig. 3–2 Sporophytic self-incompatibility. In this diagram the S-alleles are presumed to act independently. Other relationships are known, including dominance and mutual weakening. ES – embryo sac.

Since the S locus in both gametophytic and sporophytic self-incompatibility systems can have many alleles – sometimes amounting to scores – a species population will consist at any one time of numerous breeding groups, each partly or fully sterile within itself because of identity at the S locus but fertile with all others.

3.2.2   The sporophytic system

In the two principal families with sporophytic self-incompatibility systems, the Cruciferae and the Compositae, incompatible pollen grains are inhibited while still on the stigma, before the pollen tube has emerged or very soon after. The recognition is therefore very rapid, probably taking no more than a few minutes. In this time there must be mutual identification, and the pollen grain must somehow disclose the S-genotype of its parent. There must then follow a reaction, or a sequence of reactions, which prevents the pollen tube from emerging from the grain, or arrests it soon afterwards.

The recognition event apparently involves materials carried in or on the walls of the interacting partners, pollen grain on one side and stigma surface on the other (HESLOP-HARRISON, 1975 a and b). The pollen grains of the Cruciferae and Compositae are enclosed in a compound wall, with an inner pectocellulosic layer and an outer layer of the special wall material, sporopollenin, thought to be a polymer or co-polymer of carotenoids and carotenoid esters. During the growth of the inner pectocellulosic layer, the intine, proteins, including several hydrolytic enzymes, are incorporated from the outer zone of the cytoplasm of the haploid vegetative cell of the pollen grain; these are sealed into the wall as thickening progresses. The outer layer, the exine, is deeply sculptured, and during the later stages of development in the anther it receives materials transferred from the surrounding diploid nurse tissue, the tapetum. These materials include lipids and pigments and also several different proteins, including glycoproteins. At maturity, then, the compound wall carries materials derived from the diploid parent in the exine and the haploid gametophyte in the intine (Fig. 3–3).

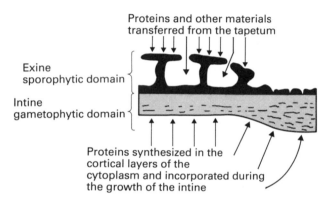

Fig. 3–3   Diagram of the pollen grain wall, showing the two domains, the exine and the intine. The exine receives materials synthesized in the diploid parental nurse tissue, the tapetum, and the intine incorporates proteins synthesized by the haploid vegetative cell of the pollen grain.

    The stigma, for its part, has a receptive surface bearing unicellular papillae in plants with the sporophytic incompatibility system. The papillae have pectocellulosic walls overlaid by a discontinuous cuticle. During development, the cells secrete materials through the wall and cuticle to form an overlying film or pellicle. This layer is mostly proteinaceous, although some lipid is present. The proteins include glycoproteins and lipoproteins, and not unexpectedly the surface binds certain lectins.

    The materials held by pollen and stigma include constituents that are antigenic in rabbits. NASRALLAH and WALLACE (1967) compared the antigens of *Brassica oleracea* (cabbage) of different S-genotypes from both pollen and stigma. No variations were found among the pollen antigens, but in one comparison differences were found among those from the stigma, and these were correlated with the S-genotype. These authors gave evidence that the antigens were proteins (possibly of course glycoproteins), good reason for believing that they might be involved in the recognition reactions of the self-incompatibility system.

When the pollen is received on the stigma it becomes hydrated, and the wall-held materials pass out onto the stigma surface where they bind with the overlying pellicle (Fig. 3–4). This happens in the first few minutes of contact, and presumably it is at this time that the recognition event occurs. Thereafter in a compatible combination the tube tip dissolves away the cuticle of the papilla enzymically, and the tube grows down into the style between the cuticle and the wall. In an incompatible combination the growth of the male gametophyte is arrested, and the tip zone becomes occluded with callose, a β-1,3-linked glucan. In the adjacent papilla, and in register with the contact face with the grain, a lenticule of callose is also formed at the plasmalemma. Both partners therefore react in a quickly detectable way after the initial recognition. The reaction of the stigma can be used as a bioassay for the presence of pollen-borne recognition factors, since callose will not be deposited unless these

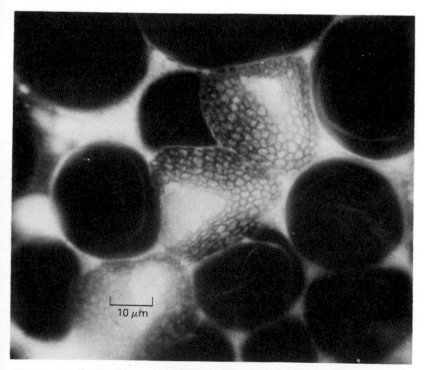

Fig. 3–4   Pollen grains of a candytuft species, *Iberis semperflorens* (Cruciferae), on the stigma; fluorescence micrograph, proteins stained with the fluorescent stain 1-anilino-naphthylsulphonic acid. The proteins held in the reticulate exine are passing out on to the surface of the stigma papillae, where they bind within a few minutes of contact.

have been received from incompatible pollen. Making use of this method, evidence has been obtained suggesting that the recognition factors are carried in the exine of the pollen grain, having been derived from the tapetum of the anther of the pollen parent.

The biological activity of the stigma factors has been tested by using pollen grown *in vitro*. When stigma extracts from a compatible genotype are added to the medium in low concentration, growth proceeds normally, but when those from an incompatible genotype are incorporated, growth is inhibited.

These experiments point to the likelihood that the *S*-gene products interact very soon after the contact between pollen grain and stigma in the primary recognition reaction, setting in train secondary responses which determine either 'acceptance' or 'rejection'. A possible scheme for the whole system is given in Fig. 3–5. It will be seen that genetical control by the diploid pollen parent, the characteristic of the sporophytic system, is readily explained if the recognition factors are derived from the anther tapetum.

The recognition factors are yet to be identified and characterized chemically. In an incompatible combination, identical *S*-alleles are present in pollen and stigma, so it might be expected that identical gene products would be present; yet the finding that antigenic differences can be detected between *S*-genotypes in the stigma but not in the pollen shows this cannot be so.

3.2.3   *The gametophytic system*

In most angiosperm families with the gametophytic self-incompatibility system the discrimination between acceptable and unacceptable pollen tubes is made while the tubes are growing through the style. Incompatible tubes grow more slowly, or are checked altogether, while compatible tubes grow through to the ovary. DE NETTANCOURT *et al.* (1973) have made a detailed study of the response in a species of the tomato genus *Lycopersicum* (Solanaceae). Here the tubes grow through a transmitting tract in the style, making their way between the cells in a matrix material. All tubes begin growth in much the same way, but incompatible ones soon show abnormal tip growth. Sometimes the wall bursts with the release of polysaccharide precursor materials, and in other cases growth simply ceases and the tip becomes occluded with callose, as in inhibited tubes in species with the sporophytic system.

Evidently, then, in the Solanaceae and families with a similar self-incompatibility control, recognition does not take place on the stigma surface. On the pollen side, the site of the inhibition suggests that materials carried in the pollen wall are not likely to be concerned; but the response would be accounted for were the recognition factors released, or exposed, later while the tube was growing through the style. On the female side, the recognition factors must be present in the matrix materials through which the tubes grow in the transmitting tract.

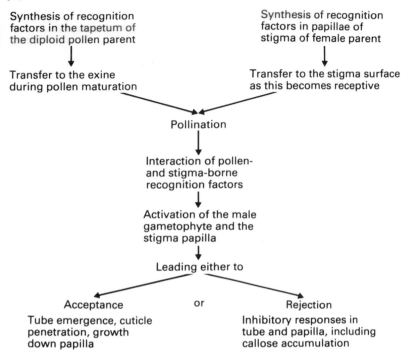

Synthesis of recognition factors in the tapetum of the diploid pollen parent

↓

Transfer to the exine during pollen maturation

Synthesis of recognition factors in papillae of stigma of female parent

↓

Transfer to the stigma surface as this becomes receptive

Pollination

↓

Interaction of pollen- and stigma-borne recognition factors

↓

Activation of the male gametophyte and the stigma papilla

↓

Leading either to

Acceptance   or   Rejection

Tube emergence, cuticle penetration, growth down papilla

Inhibitory responses in tube and papilla, including callose accumulation

Fig. 3–5 Scheme for the interaction of pollen and stigma in species with the sporophytic self-incompatibility system.

Morphologically, this occupies the site of the middle lamellae (Fig. 3–6). The material contains polysaccharides, and considerable quantities of protein, including glycoproteins. The probability that the recognition factors are held here is reinforced by the observation made some years ago that pollen tubes of *Petunia* (Solanaceae; gametophytic self-incompatibility system) growing *in vitro* are inhibited by extracts from styles bearing the same S-alleles, but not by extracts from those of different S-genotype.

In two families with the gametophytic system the inhibition of incompatible tubes is not in the style but at or near the stigma surface. In one of these, the grass family (Gramineae), the stigma has no fluid secretion, but the papillae that receive the pollen grains have a proteinaceous surface layer like that found in plants with the sporophytic system. In an incompatible match the tubes are inhibited very soon after the tips make contact with this layer (Fig. 3–7), which is likely therefore to be the recognition site. In *Oenothera organensis* (Onagraceae), a species of evening primrose, the stigma surface bears a fluid secretion when it is receptive, and the surface cells are dead or moribund. Incompatible tubes

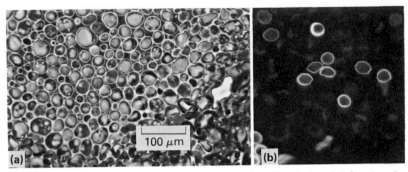

Fig. 3–6 Transverse section of the style of a species of *Malus* (apple) showing the cells of the transmitting tissue in section and the system of intercellular spaces. (a) Stained for protein; (b) stained with decolourized aniline blue and viewed with the fluorescence microscope. The pollen tubes can be seen in cross section because of the fluorescence of the inner callosic wall layer; the cells of the transmitting tissues in this species have no callosic layer and are thus barely visible.

are inhibited in this layer, showing that it is the site of the recognition reaction.

For there to be such an early response in the grasses and *Oenothera*, the recognition factors must be presented by the pollen very quickly after it is captured. These factors may be held, pre-packaged, in the intine of the grass pollen grain, since proteins from this site are released onto the stigma surface as soon as the grains are hydrated. The work of LEWIS,

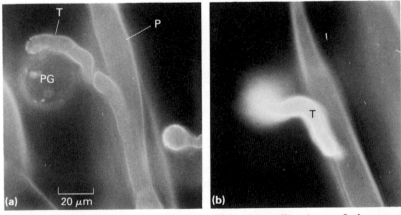

Fig. 3–7 Compatible (a) and incompatible (b) pollinations of the grass *Alopecurus pratensis*. In the compatible pollination the tube (T) has emerged from the grain (PG) and has penetrated the cuticle of the stigma papilla (P) before growing on towards the ovary. In the incompatible pollination, the tube (T) has emerged, touched the surface of the papilla, and then been arrested. Callose has accumulated both in the tube and in the grain. Preparations stained with decolourized aniline blue and viewed with the fluorescence microscope. (× 500.)

BURRAGE and WALLS (1967) suggests that the S-gene products are similarly pre-packaged in *Oenothera*. These authors found that antigens specific for the S-genotype of individual pollen grains diffuse from intact grains while they are hydrating and before there is any pollen tube growth. These antigens may be the recognition factors themselves, but this remains to be proved.

3.3 Interspecific incompatibility systems

We have noted that related species may be kept apart genetically by physiological maladjustments between pollen and stigma that can act as sterility barriers. Control of a more specific kind apparently involving recognition responses has been investigated by KNOX, WILLING and PRIOR (1973) in two willow species, *Salix alba* and *S. deltoides*. The species are normally intersterile when crossed in either direction, the foreign pollen being blocked on the stigma. In one experiment, the surface materials from samples of pollen of *S. alba* were removed, and the protein fraction separated; this was then added back to the stigmas of flowers of *S. alba* with the normally incompatible pollen of *S. deltoides*. A considerable seed set was obtained, and the progeny were hybrid. Similar results were achieved when killed pollen of *S. alba* was supplied together with that of *S. deltoides*.

Such results suggest that the recognition system of the stigma can be confused if it is saturated with 'signals' from a compatible source, even to the extent that it does not respond when incompatible pollen is received later. Interestingly, the technique of mixed pollination has been used in forestry practice to produce hybrids that cannot otherwise be obtained, and the method was also adopted extensively by Russian plant breeders to obtain remote hybrids among fruit-trees species. Compatible pollen used simultaneously with foreign pollen in this way is sometimes referred to as *mentor* pollen, to convey the idea that it somehow leads or instructs the foreign tubes in their passage to the ovary. This might be interpreted as meaning that the compatible pollen reduces the effectiveness of the surveillance system by saturating the receptor sites.

Although little is known about how interspecific incompatibility systems work, sometimes they seem to have features in common with self-incompatibility systems. In the Cruciferae, for example, the foreign pollen in many interspecific crosses is checked on the stigma with the formation of callose both in the stigma and in the arrested tubes, just as though it were from an incompatible genotype of the same species. This appears to be a very specific response since it is only expressed when the pollen comes from other Cruciferae. Pollen from other families may be stopped on the stigma, but it does not induce any response in the papillae. Experiments with the Compositae have shown that in this family, too, plants can recognize pollen from related species and genera while ignoring that from other families.

4 Interactions between Somatic Cells in Vascular Plants

4.1 Introduction

As was noted in Chapter 1, a considerable body of knowledge now exists about how animal cells communicate and otherwise interact during embryogenesis and development. As yet this is not matched by any correspondingly abundant information about short-range interactions of cells in the somatic tissues of vascular plants. This could be because they are less significant in plants than in animals, but it seems more probable that the deficiency results simply from the neglect of this field of enquiry, for the fact is that research on this aspect of the physiology and cell biology of plants is still in its infancy. In this chapter a few situations will be reviewed where somatic tissues are brought together in circumstances in which one might expect specific cell interactions to be detectable. In none can it be said that a satisfyingly complete account of events has been given; but each poses its own set of questions, and in some instances the direction that research might now take will be obvious enough. Somatic cell interactions in normal differentiation and development are considered in Chapter 6.

4.2 Artificial grafting

The age-old practice of grafting aims to bring together parts of plants of different genotype in such a manner that they will develop a tissue union and function physiologically as a single individual thereafter. The methods of producing grafts are manifold, and their details form part of the craft-lore of horticulture and arboriculture. The transferred shoot is the scion, and the portion of stem and root receiving it is the stock. The scion may be no more than a bud, inserted in a prepared crevice in the stock, or it may be a more fully developed shoot system set into the stock by matching cuts designed to give good tissue contact and mechanical strength. For success it is normally necessary to protect the site of grafting from desiccation and to prevent excessive transpiration from the scion while the vascular union is being established. A graft that 'takes' is referred to as compatible and one that fails as incompatible.

Several earlier plant anatomists investigated the structure of graft unions, but only in the last few years has the whole process been studied as a possible example of cell recognition. In a compatible union cells and tissues in the neighbourhood of the graft site must cooperate in scavenging the wound zone and in restoring the continuity of the conducting systems and of tissues such as the cambium and epidermis or bark, and this cooperation must necessarily be based on mutual acceptance. Incompatibility could arise from many causes, ranging from a general incapacity to make the necessary structural and physiological adjustments in the case of combinations between taxonomically distant species, to specific rejections even when stock and scion are from related species. LINDSAY, YEOMAN and BROWN (1975) and YEOMAN and BROWN (1976) have compared compatible and incompatible grafts in the Solanaceae with the object of clarifying the cytological and physiological reasons for the difference. The grafts investigated were all prepared by the simplest possible method, namely by cutting the stem of one seedling, the prospective stock, across transversely, and apposing to it a scion obtained by cutting a donor seedling across at a level where the stem diameter was the same. The graft sites were protected from drying out, and the scions were maintained in a humid atmosphere to reduce transpiration. To follow the course of fusion, the strength of the graft union was determined in batches of identically prepared plants at daily intervals by applying increasing tensions to the scion until it became detached from the stock. Various graft combinations were tested, including self-on-self (autografts), between plants of the same species (homografts) and between different species or genera (heterografts or allografts).

The change of breaking strength with time for different types of grafts of *Lycopersicum esculentum* and *Nicandra physalodes* are shown in Fig. 4–1. The curves for the autografts are typical for fully compatible grafts in the Solanaceae, and the curve for the allograft, *Lycopersicum* on *Nicandra*, shows what happens in an incompatible union. YEOMAN and BROWN (1976) have described the cytological events over the period of time spanned in Fig. 4–1. In both compatible and incompatible combinations, the cohesion between partners increases during the first few hours, before there could be any cell division consequent upon the wounding. Thereafter the strength of the union continues to increase over the first four days, when, in the compatible graft, tracheids are formed and the lignification of cells in the junction region begins a second phase of rapidly increasing strength until the maximum is reached three or four days later. In incompatible allografts like that of *Lycopersicum* on *Nicandra* (Fig. 4–1), the cohesion increases slowly during the week following the first contact, but true tissue union is not achieved and tracheids are not formed. In the allograft there is thus no second phase.

In both compatible and incompatible combinations, however, mitosis begins in cells on either side of the contact face during the first period. While this is happening, the tissues near the centre of the apposed stems remain in close contact, and the remnants of the damaged cells of the cut surfaces are dispersed, presumably through enzymic digestion. The cortical tissues shrink somewhat, but the gaps between the tissues of stock and scion are quickly filled by new cells while the wound debris is scavenged (Fig. 4–2). The cohesion between the apposed tissues, first in the

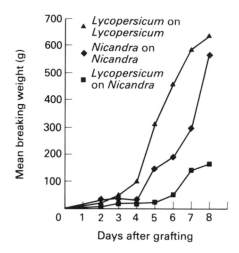

Fig. 4–1 Breaking weights of butt-grafts on successive days after grafting of *Lycopersicum esculentum* or *L. esculentum* (autograft), *Nicandra physalodes* on *N. physalodes* (autograft), and *L. esculentum* on *N. physalodes* (heterograft). (From YEOMAN and BROWN, 1976.)

central region and then in the cortex, appears to result from intensive secretory activity on the part of the undamaged cells at the interface, the secreted products including pectic substances and probably also the precursors of other wall polysaccharides. In consequence of this secretion the walls at the interface thicken substantially during the first period while the cohesion is increasing and before any lignification. The consequence of these changes is to reconstitute in the transition zone the kind of relationship seen in intact parenchyma, with apposing cell walls cemented together by the pectins and other polysaccharides of the middle lamella.

In the allograft the bonding does not become so secure, but the first major structural differences do not really become apparent until the cells destined to form vascular connections grow towards each other from stock and scion. In the compatible combinations these touch, the end walls are eroded and cytoplasmic contact is established, preparatory to the formation of vascular elements. In the incompatible combination there is a mutual rejection after the cytoplasmic contact is made and vascular elements are never produced across the union.

YEOMAN and BROWN (1976) conclude that, while the preliminaries are much the same in compatible and incompatible combinations, a critical difference develops when proteins secreted into the area of the graft union begin to interact. They suggest that in the compatible combination an activating complex is formed which promotes continued development, while no such combination arises in the incompatible. Evidence has been obtained showing that new kinds of proteins are

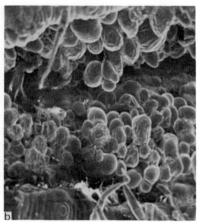

Fig. 4-2 Scanning electron micrographs of graft contact faces, seen from the side. (a) Heterograft of pepper (*Capsicum frutescens*) on stock of tomato (*Lycopersicum esculentum*), six days after grafting. (b) Autograft of tomato, four days after grafting. In each case the gap in the site of grafting is being filled by the ingrowth of new cells from the cut surfaces. In the compatible case, the autograft, cytoplasmic contacts will be made and vascular continuity established. In the incompatible, the heterograft, there will be a mutual rejection after the cytoplasmic contact. (Electron micrographs reproduced by the courtesy of Dr M. Yeoman.) (a, × 300, b, × 350.)

indeed synthesized in the vicinity of the graft union, concerned probably both with the recognition system and the scavenging and repair processes. Among the new proteins are some with the properties of lectins, and in work with autografts of another solanaceous genus, *Datura* ('thornapple'), these have been shown to increase on both sides of the graft union. The lectins are probably themselves glycoproteins; the functions they might perform remain to be determined, but they are obvious candidates for participation in the recognition reactions.

The literature dealing with the grafting of woody plants is extensive, and particularly that dealing with genera of bush and tree fruits, but no studies have been made as yet on the cytological and biochemical events at the graft union in sufficient detail to throw any light on specific recognition mechanisms that may be at work. The failure of taxo-nomically remote grafts involving different families or genera is often marked by the death of tissues in the vicinity of the contact face. Such responses might merely indicate that one or other or both partners are adversely affected by the transfer of toxic metabolites, and this can scarcely be looked upon as a specific recognition event. In other instances failure is associated with the more or less copious development of wound tissue by the cambium of the stock, leading often to the sealing of the cut surface, much as if the presence of the scion were simply being ignored. Some of the records of grafting among fruit trees suggest that tissues may vary in their readiness to accept one and the same graft partner during development, since young plants used as stocks may be more tolerant than old ones of the same genotype. This may merely reflect metabolic differences between juvenile and adult tissues, and could again be related to the production of toxic metabolites. On the other hand, it could reflect changes in the surface properties of the cells, or in their capacity to respond to the presence of foreign cells, of a much more specific kind. Such instances now require re-investigation to establish what type of communication there is across the graft site and to discover whether rejection or acceptance is associated with the secretion of special factors such as those postulated by Yeoman and Brown.

4.3   Natural grafting

Natural grafting between the aerial parts of plants is not common, but it does occur. It is a regular feature of the growth of the strangling figs and certain other epiphytes and lianas, and is found occasionally even among forest trees like the beech. Such natural grafts are autografts; homografts are rare, and allografts of arboreal species in natural communities seem not to have been recorded. To this extent, then, the situations created in artificial grafting have no natural counterpart. However, root grafting is quite common, perhaps because the less disturbed conditions of the root environment provide the stability needed for the unions to be forged. Autografting among the roots of certain species of forest trees is almost invariable, as may be verified from the inspection of any exposed root system of beech or oak (Fig. 4–3). Homografting between the roots of different individuals in pure stands of single species is also common; indeed it is likely that many forestry plantations become eventually one single 'organism', physiologically speaking, by the repeated establishment of such connections. In natural forests there is probably much less homografting of the root systems, partly because individuals are more dispersed in mixed associations, but also probably because of the greater genetical diversity in the population. It sometimes appears that the root systems of contiguous individuals in natural stands form fewer homografts than autografts, suggesting some form of discrimination. This might well be favoured by natural selection since free homografting could increase the hazard of uncontrolled spread of systemic disease.

4.4   Chimaeras

Chimaeras are organisms built up of two or more genetically different tissues, different from graft combinations in that the tissues are associated throughout the body, or a major part of it, and not only at the site of a

Fig. 4–3 Exposed root system of an old beech tree, showing repeated autografting.

graft union. They are quite common among vascular plants, and many garden plants grown for their variegated foliage are of this kind. The genetical differences between the constituent tissues may be slight, limited to one or two genes or to a defect in the chloroplast genome; or they may be substantial when, for example, the tissues combined have been derived from different species or genera. The special feature of angiosperm chimaeras is that they often show considerable stability, due to the mode of growth through the action of apical meristems. In each meristem reasonably well-defined zones of dividing cells give rise to the various tissues of the shoot itself and of its lateral organs; if these zones or histogens are genetically different, then so will be the tissues they produce. A common form of chimaera has the inner part of the meristem composed of the cells of one partner and the outer sheath of those of another (Fig. 4–4). Such periclinal chimaeras are often very stable, and since all organs of the plant have the same disposition of tissues the fact that it is chimaerical may not be easily detected. In another version, a sector of the apical meristem may be of one tissue and the remainder of another. In this case the lateral organs, leaves and branches, will differ according to the sector from which they arise.

Chimaeras arise spontaneously after grafting, when adventitious shoots are produced in the vicinity of the graft union containing tissues of both stock and scion, or they may be encouraged to form by cutting across the graft union to promote the formation of mixed callus and regenerating shoots. In this way many species combinations have been produced in both herbaceous and woody plants. '*Solanum tubigense*' is a chimaera with one cell layer of tomato (*Lycopersicum esculentum*) overlying a

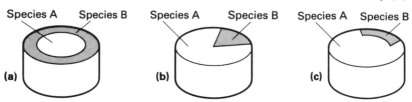

Fig. 4–4   Types of chimaera. The stem sections show the distribution of the two partner tissues in (a) a periclinal chimaera, (b) a sectorial chimaera, and (c) a mericlinal chimaera.

core of *Solanum nigrum*, and '*Solanum koelreuterianum*' is the reciprocal, with *S. nigrum* overlying tomato. '*Cytisus adami*' is a chimaera of purple broom, *Cytisus purpurea*, overlying laburnum, *Laburnum vulgare*.

The fact that interspecific and intergeneric chimaeras exist, and show the stability they do, proves a very significant point, evident also from the work on grafting, namely that the appearance of somatic incompatibility is not a *necessary* accompaniment of evolutionary divergence in plants. It is inconceivable that tissues should cooperate so intimately in development if there were such barriers. Furthermore, viewing the matter in another way it can be seen that for such cooperation to be possible the participating tissues must have quite similar hormonal responses and must react in very much the same way to such short-range communication systems as might be involved in the control of morphogenesis. Coordination is not always perfect, however, as shown by the appearance of distorted leaf forms and other abnormalities in chimaeras between tissues of plants with different leaf shapes. The retention of cellular identity is sometimes expressed in other ways also; in *Solanum tubigense* the tissues derived from the tomato are susceptible to attack by the fungus *Septoria lycopersici*, but not those from *Solanum nigrum*.

4.5   Cells in culture

Much of the knowledge of the interactions of animal cells has been gained from work on tissues and isolated cells growing in artificial culture after explanting from the parent organism. Although plant cells and tissues can readily be grown in culture, they do not lend themselves to the same kinds of experimentation because they grow in a different manner and, being encased in walls and incapable of movement, do not interact so freely. Nevertheless a certain amount of evidence has been gained from observations of cell behaviour in mixed cultures.

Adjacent callus tissues grown from the same individual or from individuals of the same species are generally able to fuse and even interpenetrate each other, and such tests as have been made show that this is also true for species of the same genus. In experiments with woody fruit trees of the Rosaceae the tissues of different genera have been found to grow freely together without any evidence of rejection, a result perhaps predictable from the readiness with which intergeneric grafts can be made in this family. In other intergeneric and in many interfamilial mixtures inhibitory effects are commonly found, usually with the suppression of growth in one or other partner and the death of cells in the contact zones. However, in one or two instances cells of plants from different families have been found to grow readily in close apposition in callus cultures, even forming chimaerical tissues.

The work on cultured cells and tissues has so far given little direct evidence of specific recognition responses. Some of the inhibitory effects reported are no doubt attributable to the release of toxic products into the medium by one or other partner, and this may have little to do with specific cell interactions, as we noted for toxic effects in grafting. Nevertheless, there is much still to be done with mixed cell cultures. Systematic comparisons of the behaviour of tissues of stock and scion grown together *in vitro* with activities at the graft union in compatible and incompatible combinations could be highly informative.

4.6 Isolated protoplasts

Living protoplasts can be released from plant cells by digesting the wall with mixtures of carbohydrases which do not attack the lipids and proteins of the plasmalemma. The free protoplasts can be maintained in liquid culture in a healthy state, continuing normal metabolic activity. One of their earliest preoccupations is to begin the synthesis of a new wall, very quickly in the case of tobacco (*Nicotiana tabacum*) protoplasts, or more leisurely as in species like the vine (*Vitis vinifera*). The microfibrils of the wall are produced at the plasmalemma, where the enzymes responsible for synthesizing the cellulose chains are held.

Using wall-free plant protoplasts the properties of the plasmalemma can be explored directly, much as with animal cells. The capacity for associating with other cells can be tested, for example, and also the binding powers for marker molecules in the medium. Unfortunately the results of such experiments are ambiguous, because the surface left by the enzymic dissolution of the wall is an artefact. The distribution and configuration of the carbohydrate groups left on the surface of the plasmalemma are as much likely to be an expression of the digestion procedure used as they are to reflect any inherent properties of the plasmalemma, and so agglutination or lectin-binding properties may have little or no biological meaning unless they can be shown to be correlated with other properties of the intact cells or tissues.

Whatever their interpretation might be, differences have been found in the agglutinability of naked protoplasts. Con A agglutinates those that have been tested so far, indicating the presence of mannose and glucose residues, likely enough considering the chemical nature of the wall stripped from the protoplasts. Other lectins with different sugar specificities do not agglutinate, indicating that the particular sugars are absent or not accessible. Differences in the distribution of binding sites occur; for example, con A binds uniformly and densely to freshly prepared tobacco protoplasts (Fig. 4–5) but in scattered islets on those of

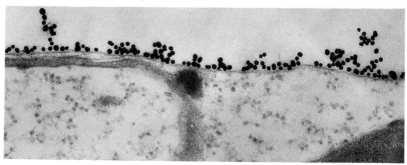

Fig. 4–5 Transmission electron micrograph showing the binding of con A coupled to colloidal gold on the surface membrane of an isolated protoplast from the tobacco leaf. The dark spherical particles, the gold-con A complexes, are dispersed along the membrane, or held in small clusters on it. In the underlying cytoplasm ribosomes may be seen, and profiles of endoplasmic reticulum. (Micrograph reproduced by courtesy of Dr J. Burgess.) (Magnification, x 40 000.)

vine. Finally, there have been some comparisons of the capacity of seed lectins to agglutinate protoplasts from the same and different species. In one instance some evidence has been obtained suggesting that mutual agglutinability might provide a guide to graft compatibility, but the principle has yet to be established as a general one.

4·7 Protoplast fusion

One purpose for which isolated protoplasts can be used is to produce somatic hybrids. By suitable manipulation using media containing high molecular weight polyethylene glycol, the cells of different species, genera and even families can be brought together and caused to fuse by the dissolution of the contiguous plasmalemmas. Results already obtained with this technique point to the likelihood that the incompatibilities revealed in orthodox grafting may not be significant once this kind of contact has been established. Thus protoplasts of cells from families as remote as the legumes and the grasses have been brought together in fusion products without immediately adverse effects. Such cells have not yet been taken through division cycles, but they do survive for some time. An interesting comparison may be made with results obtained with the fusion of animal cells, where it is clear that the histocompatibility system does not govern events once the cytoplasms have become confluent. The incompatibility signalling system is held on the cell surface, and once it is bypassed there can be no rejection.

5 Interactions in Symbiosis, Parasitism and Disease

5.1 Introduction

The cells and tissues of one organism come into close association with those of another in epiphytism, symbiosis and parasitism. The organisms may or may not bear relationship in the evolutionary sense; thus vascular plants, for example, are hosts to viruses, bacteria and fungi as symbionts or pathogens, and also accommodate other vascular plants as epiphytes or parasites. Whatever the association, there is usally some degree of specificity, either in host selection or epiphytism and parasitism, or in the choice of partner in symbiosis. This specificity implies that discriminations are made when the potential associates encounter each other, and this in turn leads us to infer that recognition systems must be at work at some point in the developing relationship.

The diversity of the types of association (Table 5) shows that the relationships must be of more than one kind. The host range of some epiphytes is quite wide, pointing to relatively low specificity. At the other end of the scale, the relationships between host and pathogen may be so intimately governed as to match in specificity the relationship between gametes of the same species. It is associations of the latter kind that are of greatest interest for the general theme of cellular recognition systems.

5.2 Host specificity and nodulating bacteria

The well known capacity of plants of the Leguminosae for fixing atmospheric nitrogen depends on a symbiotic association with bacteria of the genus *Rhizobium*. The bacteria invade the root through the root hairs and promote the formation of nodules by inducing cell division and various metabolic changes in the host tissues. The specificity of the relationship is quite high. Each legume species has its characteristic *Rhizobium*, and even within species strain differences may be found. BOHLOOL and SCHMIDT (1974) have studied the relationship between 48 different *Rhizobium* strains and the soya bean (*Glycine max*) as a host, to test the possibility that the plant lectins might be concerned in determining specificity. They found that a lectin present in soya bean bound to 22 of the 25 strains of *Rhizobium* capable of forming nodules in this species, but not to any of 23 strains tested that did not form nodules. The lectin presumably discriminates between the bacterial strains according to their surface carbohydrates, and the high correlation with infectivity strongly suggests that this is the factor determining the specificity of the

Table 5 Associations between living plants showing different degrees of intimacy and specificity.

Type of association	Specificity	Examples
Hemiepiphytism Epiphyte is independently rooted, at least initially, but gains support from the host	Low; but some lianes do seem to show host choice	Ivy; virginia creeper
Epiphytism Epiphyte gains support, protection and sometimes favourable microclimate from the association	Low to moderate host specificity	Many vascular epiphytes, including ferns, bromeliads and orchids
	Moderate to high host specificity	Many associations between unicellular and multicellular algae, and between algae, mosses and ferns and woody vascular plants
	High host specificity	Associations between various groups of marine algae, and between bryophytes and vascular plant hosts
Symbiosis Partners both gain benefit from the association	High or very high	Algae and fungi in the lichen symbiosis. Legumes and other angiosperms in association with nitrogen-fixing bacteria and blue-green algae. Angiosperm root systems in association with mycorrhizal fungi
Hemiparasitism Parasite is autotrophic, but is partly supported by nutrients acquired from the host	Moderate to high host specificity	Root hemiparasites including species of Schrophulariaceae, Santalaceae. Stem hemiparasites, including species of Loranthaceae (mistletoes)
Parasitism *Ectoparasitism* Parasite remains mainly on the exterior of the host	Moderate to high host specificity	Many vascular plant parasites, including dodders (*Cuscuta* spp.), toothworts (*Lathrea* spp.), and species of Orobanchaceae, Balano-phoraceae, Rafflesiaceae
Endoparasitism Parasites develop mainly within the tissues of the host	High to very high host specificity	Many viral, bacterial and fungal pathogens attacking other plant groups from algae and fungi to angio-sperms

relationship and indicates furthermore that the lectin may play some part in the discriminations. This study has recently been extended to other species of legumes, and it has been found that each species tested produces a lectin that binds specifically to its appropriate bacterial symbionts. The binding seems to be with specific sugar groups or sequences in the lipopolysaccharides of the bacterial surface.

These findings have important implications. They provide a model for comparison with other examples of powerful host specificity, including those found in disease, and they also offer a clue as to the possible function of lectins in the plant body, a matter discussed further in section 5.6.

5.3   Host specificity in fungal disease

The relationship between host and pathogen is always a subtle one, both in the interactions within the infected individual and in the balance between the populations of each in the wild or in cultivation. An incompetent pathogen risks extinction; but so does a too aggressive one if it succeeds in destroying its hosts. It is not surprising therefore that the co-evolution of host and pathogen should often have led to the development of quite sophisticated systems of cellular communication and interaction.

Fungal diseases in flowering plants provide excellent examples of this. Plant species vary greatly in their susceptibility to disease, and this is true also for varieties and cultivars. On the other side, fungal species differ in their host ranges, and different strains or races of a pathogen can vary in their capacity for attacking one and the same host. Host defence against fungal attack takes many forms, but it is sometimes useful to make a distinction between generalized and specific resistance. Generalized resistance is the capacity to deal with fungal invasion by blanket measures, effective against a wide range of potential pathogens. Specific resistance describes the ability to mount a defence against particular pathogens, or against individual races or strains of a pathogen. Potatoes show race-specific resistance to different genotypes of blight, *Phytophthora infestans*, and various cereal species to races of rusts (*Puccinia* spp.)

In these and many other instances resistance on the part of the host and pathogenicity on the part of the fungus have proved to be connected in a rather precise way. Resistance in the host variety to a pathogen race may depend on a particular gene, the functioning of which inhibits the attack of the race for which it is specific. Races to which the variety is susceptible differ genetically in such a way that they are not inhibited by the gene giving resistance. A possible interpretation of this situation favoured by some plant pathologists is that the natural state of the host species is susceptibility, the capacity for resistance arising when genes are present that confer the ability to recognize pathogen races and initiate defence measures against them. A mutation in the fungus that prevented this recognition would produce a new virulent race, and this would be successful as a pathogen until such a time as the host produced a counterpart mutation permitting recognition once again. In this way a gene-for-gene relationship would be established between pathogen races and host varieties (FLOR, 1971).

This integration of race-specific resistance attributes a well-defined but limited role to the resistance genes of the host, namely that of coding for recognition factors which act, so to speak, as the sentries guarding host tissues. The blockage of the pathogen is the responsibility of other systems. This view is supported by studies of the nature of the host's defences. Race-specific resistance usually results from so-called hyper-sensitive reactions in the host (reviewed by DAY, 1974). Cells suffering the invasion and those in the immediate vicinity undergo metabolic changes, become necrotic and then release substances, the phytoalexins, that inhibit the growth of the pathogen. The invasion of a pathogen race for which the host does not have resistance genes does not initiate this response: the host behaves as though it were unaware of its presence. In such a circumstance it might be expected that a simultaneous invasion by virulent and non-virulent races would lead to some reaction against both, since the non-virulent race would activate the alarm system, even if the virulent race did not. Experiments with wheat rusts show the expectation to be justified (Table 6). These experiments suggest that the host's control

Table 6   Spore production on seedling leaves of the wheat variety Maris Beacon inoculated with *Puccinia striiformis*, virulent race (V) only, and with virulent and non-virulent races (V + NV). In one experiment inoculation with the virulent race was delayed for four days. (Data from JOHNSON, 1976.)

Leaf surfaces treated	Days between NV and V	Spore production (mg cm^{-2} leaf) V	NV + V
Opposite	0	115.1	63.0
Upper	0	122.2	53.2
Upper	4	102.1	49.6

measures are themselves not specific, the specificity lying in the recognition step.

What, then, is the basis of the recognition? ALBERSHEIM and ANDERSON-PROUTY (1975) have proposed a scheme based on the idea that the pathogen carries surface molecules which act as identifying labels. If the potential host has receptors capable of recognizing these, then the defence system can be alerted and the host will be resistant. If, on the other hand, receptors for the pathogen markers are lacking, no defence will be set up and the host will be susceptible. Pointing to the analogy with other cellular recognition systems, including those found in animals, Albersheim and Anderson-Prouty suggest that the surface factors carried by the pathogen are glycoproteins, the specificity lying in the carbohydrate part of the molecules. The recognition receptors in the host would then be the complementary proteins, held in the plasmalemma, or near to it, in or on the wall. Mutation in the pathogen, presumably involving a change in the specificity of a glycosyltransferase, would produce a new coding of the surface polysaccharides; the host would no longer have receptors for this, and so would be susceptible. A mutation of the gene coding for the receptor which permitted recognition of the new sugar configurations on the cells of the pathogen would restore resistance. The scheme is summarized in Fig. 5–1.

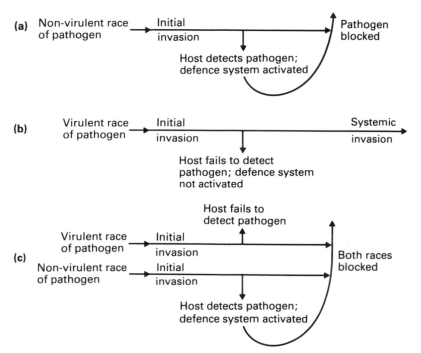

Fig. 5–1   Host responses to the invasion of (a) a non-virulent pathogen, (b) a virulent pathogen, and (c) virulent and non-virulent races together.

5.4   Inducers and receptors

The scheme of Fig. 5–1 is as yet hypothetical, and the glycoproteins and receptors for them envisaged by Albersheim and Anderson-Prouty have not been isolated and characterized in any instance of a host-parasite interaction. However, the authors point to various lines of supporting evidence. It has been found that certain fungal pathogens in culture release substances into the medium that can induce the production of phytoalexins when applied to the tissues of resistant host varieties, and these substances are glycopeptides. In one such example, Albersheim and collaborators have found furthermore that very active inducing substances can be derived from purified mycelial walls. The effectiveness of these appears to depend on the detailed structure of a polysaccharide component. These results strongly suggest that the host does indeed recognize the pathogen by the configuration of carbohydrates on the surface of the hyphae. Then the fact that the cells of plants do have protein receptor molecules in the cell membrane capable of binding glycosides produced by pathogens has been demonstrated by STROBEL and HESS (1974), in work on *Helminthosporium sacchari*, a fungus causing eyespot disease in sugar cane. This fungus produces a toxin that is only active against its specific host. The toxin is a small glycoside, and this binds to a membrane protein present in susceptible but not resistant varieties of the host. In this instance the binding of the toxin in susceptible varieties leads to changes in the cells and symptoms of the disease. In the scheme of Fig. 5–1 it is envisaged that binding initiates the defence reaction against the pathogen.

5.5   Host-pathogen and self-incompatibility systems

The similarity between the interactions in race specific pathogenicity and those in angiosperm self-incompatibility systems is very striking. In each case, the tissues of the plant are penetrated by a filamentous invader, hyphae in the one, pollen tubes in the other. In each, the invader is either accepted (susceptibility or compatibility) or rejected (resistance or incompatibility). The control in both instances is by one or a few gene loci in each of the interactants. The recognition events seem to depend on surface properties, in that the invading cell or cells carries specific signals for which the host has receptors, at least where the response is to be rejection. Finally, there is some indication of a common chemical basis for the interactions. At this time little more can be said in pursuance of the analogy since so much of the detail in each situation remains to be studied. It is an intriguing thought, nevertheless, that perhaps in the course of the evolution of the flowering plants, mechanisms first useful for combating disease were later adapted for the regulation of the breeding system by controlling the growth of pollen tubes.

5.6   Non-specific disease resistance

Any measure taken by a host against an invasive pathogen must be specific to the extent that it discriminates against foreign cells while preserving the health of host tissues. So, even generalized disease resistance must involve some tissue recognition mechanism, however crude this might be. A defensive strategy based on the identification of some characteristic chemical or other feature of a class of pathogens might be expected to provide protection against the whole of the class. The first, and still tentative, evidence that flowering plants may have systems of this sort for dealing with fungi has now been obtained by MIRELMAN, GALUN, SHARON and LOTAN (1975). A lectin from the wheat (*Triticum aestivum*) embryo (wheat germ agglutinin, WGA) binds specifically to saccharides containing N-acetylglucosamine. Chitin, a characteristic constituent of some fungal walls, is a polymer of this amino sugar. Mirelman *et al.* found that WGA binds to the growing hyphal tips of the fungus *Trichoderma viride*, at zones where the chitin is exposed and apparently not yet protected by glucans (Fig. 5–2). The binding brings hyphal

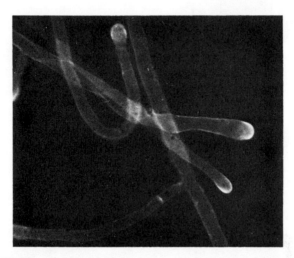

Fig. 5–2 Binding of the lectin, wheat germ agglutinin, to the growing hyphal tips of the fungus *Trichoderma viride*. The lectin has been labelled with a fluorescent tracer, and this micrograph is of the mycelium illuminated with ultraviolet light. The sites of binding are seen as the zones of brighter fluorescence. The lectin binds to saccharides containing N-acetylglucosamine, the amino sugar present in chitin, a characteristic constituent of some fungal cell walls. (Micrograph reproduced by the courtesy of Dr E. Galun.) (Magnification, × 700+.)

growth to a stop, perhaps by prematurely cross-linking extending chitin chains. So, in this situation, WGA acts as a fungistatic agent. The authors suggest that the lectin may function in the same way during germination of the wheat grain to protect the seedling against invading fungal pathogens. The proposal is attractive, but proof that the lectin performs this role in the natural situation is still to be given.

5.7   Epiphytism and ectoparasitism

No examples of specific association between epiphytes and ectoparasites and their hosts have been examined as yet in physiological detail, so little or nothing can be said about the biochemistry of the recognition systems that might be at work. The circumstantial evidence indicating that these must exist is, however, overwhelmingly strong.

Algae freely form epiphytic associations, unicells on filaments, filaments on the more massive forms, and the larger algae among themselves (Fig. 5–3a). In some instances the relationships seem to be

Fig. 5–3 (a) Algal epiphytism. The epiphyte is *Polysiphonia fastigiata* (Rhodophyceae) and the host *Ascophyllum nodosum* (Phaeophyceae). Species of *Polysiphonia* are obligate epiphytes on species of the Fucales. (b) Mistletoes (*Viscum album*) growing on the branches of apple (*Malus domestica*). *Viscum album* is a hemiparasite occurring on a limited range of tree species in Britain.

almost obligatory; the characteristic habitats of certain marine red algae (Rhodophyceae), for example, are the surfaces of other plants, and these are occupied with great fidelity. Relationships like this may have quite simple explanations, notwithstanding the specificity; the host might be the preferred habitat just because it happens to pass desirable nutrients into the medium. There would be no need in such circumstances to postulate the existence of very precise recognition mechanisms. Nevertheless, the possibility that the specificity does result from special kinds of molecular interaction remains open. Some algae attach themselves to the substratum by secreting viscous mucopolysaccharides which bond tenaciously even under sea-water. Conceivably the adhesion of the sporelings of some epiphytes, the essential preliminary to establishment, can only be achieved on host surfaces of a suitable chemical nature.

Vascular-plant parasites and hemiparasites establish very close connections with their hosts. Familiar examples are the broomrapes (*Orobanche* spp.) and dodders (*Cuscuta* spp.), both chlorophyll-deficient parasites, and the eyebrights (*Euphrasia* spp.) and mistletoes (family Loranthaceae; Fig. 5–3b), root and stem hemiparasites. Here again there are varying degrees of host specificity. The relationship is very often a species-for-species one, each host with its characteristic parasitic or hemiparasitic associate and each parasite or hemiparasite with its characteristic host, but in some instances there may be racial or varietal specificity.

Several root parasites have evolved methods of detecting the presence of potential hosts. Good examples are provided by species of *Striga* (witchweeds: Scrophulariaceae). These are important parasitic weeds of cultivation in tropical countries, sometimes accounting for severe losses in crops of maize (*Zea mays*) and sorghum (*Sorghum vulgare*). The seeds of the parasite germinate best in the vicinity of the host roots, and this is due to the presence of promoting substances in the root exudates. One of these, strigol, a highly potent factor from cotton (*Gossypium* spp.) roots, has been isolated and characterized. The presence of such exudates evidently indicates to the seeds of the parasite that actively growing hosts are available in the vicinity. Furthermore, the behaviour of the parasite is affected in other ways by the root exudates after germination. The emerging radicle grows towards the root in a chemotropic response, and at closer range morphological changes are induced in the seedling preparatory to the penetration of the host tissues. Although these responses are certainly important for the parasite in locating the host and making the first contact, it is improbable that they account for host specificity, since promoting factors have been found in exudates of roots of plants that are not normally hosts at all. The actual selection of the host must follow contact, and the apparent 'choice' probably relates mainly to how easily the parasite can make its first penetration and begin growth through the tissues. On the host side, it is at this time that any specific protection mechanisms might be expected to act, if the analogy with fungal disease is valid. Stem parasites like the dodders must rely on identifying the host after contact with its living aerial stems, and these contacts are made by wide nutatory movements of the stem of the seedling parasite, without, so far as we know, any guidance system.

6 The Operation of Cellular Recognition Systems: Fact and Speculation

6.1 How many systems?

The examples of short-range intercellular communication discussed in the foregoing chapters illustrate a variety of situations, and we must now seek to relate these to each other. The fullest information comes from the examples of interaction between independent cells or organisms. In these the response may: (a) depend on the receipt of diffusable signals from one cell or tissue by the receptors of another cell or tissue (e.g. aggregation in the slime moulds; pheromones in algae and fungi); (b) follow cell wall contacts (e.g. conjugation in yeasts; incompatibility systems in flowering plants); (c) follow contact between cell membranes (e.g. gamete fusion in *Chlamydomonas* and *Fucus*; morphogenesis in slime moulds); (d) require cytoplasmic contacts (e.g. plasmogamy in Basidiomycetes).

The sequence (a) to (d) marks increasingly close interaction, with the potential, at least, of progressively greater specificity. As we have seen, some of the more complex interactions involve communication in more than one mode, with the 'right' result being obtained only when all parts of the system function.

At this point we can see why it is not feasible to define cellular recognition any more precisely than we attempted at the outset, and why it is preferable rather to accept the concept as a broad one covering all kinds of short-range cellular communication which elicit specific responses from one or both partners in the dialogue. Cells can express selectivity through each one of the interactions (a) to (d), say, in the choice of partners for a sexual fusion, and this is as far as it would be reasonable to press the idea of recognition in the present context.

We may note also that there is no difference in principle between short-range communication among cells of the kind listed under (a), above, and longer range hormonal communication in the plant body. Pheromones, after all, are simply hormones that provide for chemical communication between separate individuals rather than between different organs of the same individual. The more important question concerns the amount of information the communicating signal carries, which is another way of describing its capacity for eliciting specific reactions. The function of trisporic acid in the induction of outgrowths from the filaments of *Mucor* is comparable with that of β-indolyl acetic acid in inducing rooting or other morphogenetic responses in the stem of a flowering plant. But this is probably not true for the male-inducing substance of *Volvox*, or the sex-inducing factors of the yeasts. In these cases the communicating agents are peptides, and the specificity of the effects they produce suggests that they carry a good deal of information, coded in the amino acid sequence of the molecules, and decodable only by the appropriate receptors at the receiving end. When we consider communication at still closer range where cell contacts are involved, (b) and (c) of the list above, there is little doubt that the phenomena do depend on this kind of precise chemical interaction.

The preferential *association* of cells illustrates one aspect of this. Specific adhesions, according to our argument, are simply a sub-set of cellular recognition phenomena in genera; but they are important, and, furthermore, ideas of how they are actually managed have come from the work on animal cells. We must therefore consider them in more detail.

6.2   Recognition and adhesion

In many of the examples we have examined, the interacting cells begin their relationship by sticking together. This is true of the sex cell contacts of algae and fungi, the pollen-stigma interaction in the flowering plants, and of associations like that of legumes and their bacterial symbionts, stock and scion in graft unions, and vascular plant parasites and their hosts. These adhesions can be highly specific, and a general scheme can now be drawn up for the way they might work, a scheme with satisfactory credentials for application both to plants and animals. The requirement, of course is that the interacting cells should not merely have a general property of tackiness, but that certain cells should stick only to, or more firmly to, certain others. This can be envisaged as resulting from the existence of complementary attachment points on the surfaces of the cells, in either unipolar or bipolar arrays (Fig. 6–1). With animal cells and the naked parts of plant cells, these are likely to be held in or on the cell membranes, and with walled plant cells they must occupy external sites.

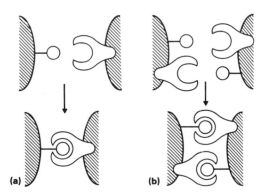

Fig. 6–1   Specific contact adhesion of two cell surfaces assuming (a) unipolar complementarity, and (b) bipolar complementarity.

The lock-and-key relationship this implies could very well be provided by the existence of lectin-like sugar-binding molecules on the one side and glycoproteins bearing the appropriate sugars or sugar sequences on the other. Such a system, in theory, could allow all the specificity ever likely to be required, and it is now favoured as an explanation of the specificity of animal cell adhesions (see review by ROTH, 1973).

Enormous diversity is possible in the heterosaccharide portion of glycoprotein and glycolipid molecules through variation in chain length, permutation of the sugar sequences, and differences in the pattern of branching. Estimates have been made of the extent of the possible diversity: for example, a chain with three different residues allows for more than 1000 forms.

If the polysaccharide portion of the glycoprotein molecules is the key, in order for the system to operate, the receptor, the lock, must have matching diversity. The favoured view is that this would be provided in the structure of proteins, although it is not excluded that the partners could be other carbohydrates. We have already encountered plant examples of complementarity between carbohydrate on the one side and proteins on the other in the cellular adhesions of the slime moulds, and of course WIESE'S work on *Chlamydomonas* (1974) points to a similar kind of mechanism, in this case governing the adhesion of the mating types by their flagella.

A molecular model has been proposed for the interaction in animal cells by Roseman, and this is reviewed by ROTH (1973). This model proposes that the receptors for the glycoprotein heterosaccharides are membrane-held glycosyltransferases, enzymes acting here in another role, but still expressing their substrate specificity. The hypothesis has the gratifying quality of reducing the number of elements in the system to which specificity has to be attributed, for the glycosyltransferases forming the lock could, acting in another mode, stamp their specificity on the key.

Whatever the chemistry of the system, it is certain that the information conveyed in, or passed through, it must originate at the gene and be expressed at one stage in the structure of proteins. Random variation in the heterosaccharides of the cell could be attributable to the presence of incompleted chains or to the unplanned incorporation of sugar residues that happen to be available, but the specificity demanded by any system where information is held in the carbohydrates can, presumably, only be imposed at the time of synthesis in the Golgi apparatus by the enzymes concerned. These are the glycosyltransferases which would, then, be the agents through which the gene-determined coding would be incorporated. The merit of Roseman's hypothesis is that it allows for the same agents to be involved in the decoding.

6.3 Cells not in contact

In some situations the selective attachment of cells determined by complementary carbohydrates and proteins may be sufficient to account for recognition, and many animal embryologists accept that this might be the basis for the sorting out of cell types in the developing embryo. By programming the cell surface factors the differentiating cells can, so to speak, determine which others they will associate with. The principle can easily be extended to account for recognitions when the cells are not actually in physical contact. Thus the heterosaccharides on their own might act as specific pheromones, diffusing to the complementary receptors on other cell surfaces. A model for plants here might be the toxin/toxin-receptor system in the eyespot disease of sugar cane (*Saccharum officinarum*). The opposite arrangement can be envisaged, with the lectin-like protein diffusible and the heterosaccharides forming part of cell-bound receptors.

For walled plant cells, the relationships might be between a diffusing protein and specific carbohydrate configurations of the cell wall. This can be invisaged as a means of controlling fungal invasion, as proposed for the *N*-acetylglucosamine-binding lectin of the wheat embryo, or possibly for the regulation of pollen-tube growth in self-incompatibility systems.

All schemes that attribute the specificity of a recognition event to interactions involving wall- or plasmalemma-bound receptors assume other links in the chain, namely one or more further reactions which determine what the cell shall do in consequence of the recognition. We have noted examples of this: the mitogenic effect of lectins in certain animal cells is a prime example, and in plants the defensive responses of tissues to fungal invasion provide another. It would be beyond the scope of this text to discuss these secondary systems.

6.4   Interactions in development and differentiation

It can hardly be doubted that the cells of multicellular plants do interact in very specific ways in the normal course of development, but just how they do this remains largely a matter for speculation. One difficulty is that adequate experimental approaches are still lacking. While the interactions between cells of different organisms can be studied with reasonable facility, it is much more difficult to do this in growing and developing tissues since the very act of interfering experimentally can disrupt the system. The problem is compounded when the aim is to study very local interactions. It has been possible to investigate the hormonal functions of auxins, gibberellins and cytokinins because the plant can be dismembered so that the flow of materials between different organs can be studied and the responses of isolated organs to possible hormonal agents tested. This is scarcely possible when the agents are transferred between contiguous cells, particularly when the movement is through plasmodesmata or from plasmalemma to plasmalemma across a distance of no more than the thickness of the intervening walls (Fig. 1–4).

One example will suffice to illustrate the kind of evidence available

from flowering plants for the occurrence of specific shortrange interactions among differentiating cells: namely the differentiation of the
stomatal complex of the grass leaf. This event involves a small group of
epidermal cells in the extending region of the leaf, and the first steps are
bound up with the determination of division planes. The epidermis is
made up of files of elongated, pavement-like cells, certain of which
become committed to forming rows of stomata. The first sign of the
differentiation is a division in an epidermal cell which produces one short
and one long daughter cell. The nuclei of the adjoining cells on either side
then migrate until they lie opposite the shorter cell, and undergo division
themselves. Cell walls are then formed to produce two subsidiary cells. A
mitotic division then follows in the short cell with the axis at right angles
to that of the file, the wall subsequently laid down dividing the epidermal
cell transversely. In this way the stomatal guard cells and two subsidiary
cells are produced (Fig. 6–2). The various cells of the complex then form
their characteristic wall thickenings. During the first period of cell
division, should one adjoining cell overlap the guard cell initials of two
stomata, the nucleus moves first to one and divides to give its subsidiary
cell, and then to the other for another division to give the second
subsidiary cell. Each subsidiary cell is in exact register with the neighbouring guard cell initial.

The briefest consideration of these events is sufficient to convince one
that such precise orientations and movements are not likely to be
controlled by freely diffusible molecules such as the auxins. To use a
term employed earlier, the cells are somehow acquiring positional
information, and are performing special acts of division and differentiation in precise coordination with each other. These activities
involve the movement of nuclei, and also, as electron microscopic studies
have shown, migration of other structures in the cytoplasm. Perhaps we
may compare these coordinated differentiations, here in one and the
same tissue, with those of neighbouring filaments of *Spirogyra* during
conjugation. The most probable explanation of the behaviour of the cells
in the differentiation of the stomatal complex is that they are responding
to inducing signals passed from one to the other. We must suppose that
the inducers are of a rather specific kind, although at present there is no
indication of their nature.

The differentiation of the stomatal complex is instructive because one
can envisage – in principle, anyway – how it might be governed by
exchange of information between the cells, bearing in mind some of
the model systems among lower plants. In the more complicated
circumstance of the apical meristem or the developing leaf primordium
where new cells are continually being added in the course of extension
growth it is more difficult to visualize how the close-range interactions
might be managed. Yet there is good experimental evidence that
communication between cells and tissues in the growing shoot apex is
essential for normal differentiation. If the young leaf primordia are

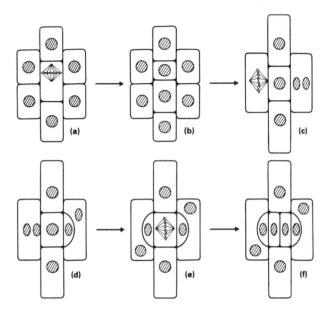

Fig. 6–2 Development of the stomatal complex of the grass leaf: (a) to (b), division giving a short and a long cell, the former the parent of the guard cells; (c) orientation of the nuclei of the flanking cells and mitosis with an axis at right angles to the long axis of the leaf; (d) formation of subsidiary cells by curved walls in register with the guard-cell mother cell; (e) mitosis in the mother cell, with the axis at right angles to the length of the leaf; (f) complete cell complex before the onset of wall thickening.

isolated from each other and from the growing point by surgical incisions in such a way that the pathways for the movement of substances between them are interrupted without blocking the flow of nutrients from the stem, they fail to follow the normal pathway of differentiation. If young enough, they adopt a radially symmetrical pattern of growth instead of the dorsiventral mode they would normally show. Evidently the primordia need to be 'informed' of the spatial relationships with each other and the growing point, in the same way that the dividing and differentiating cells of the stomatal complex require positional information to execute their particular functions correctly.

Finally, the examples from the lower plants and from the behaviour of gametes may provide us with some insight into another very characteristic feature of development and differentiation in the vascular plants. In their responses to hormones, cells frequently show that they vary in their readiness to react, behaving as though they have to be 'pre-programmed' to pick up signals and respond to them. For example, the cells of the oat (*Avena*) coleoptile, a favourite test object for growth substance assays, acquire their sensitivity during a particular period of their life, and are only competent to respond to auxins during this period. The beautiful control system of *Chlamydomonas* shows us how this may be managed. When the *Chlamydomonas* cell embarks on its career as a gamete, it signals its readiness to mate by unrolling a new area of cell membrane and exposing new types of receptor. Internal controls – in *Chlamydomonas*, related to nutrition – determine this, and these controls have nothing to do directly with those called into action later, during the conjugation process itself. Perhaps one of the important functions of close-range cellular interactions in multicellular plants is to create states of competence by changing the patterns of hormone receptors in the membranes of the cell.

References

ALBERSHEIM, P. and ANDERSON-PROUTY, A. J. (1975). Carbohydrate, proteins, cell surfaces and the biochemistry of pathogenesis. *Ann. Rev. Plant Physiol.*, **26**, 31–52.

BOHLOOL, B. B. and SCHMIDT, E. L. (1974). Lectins: a possible basis for specificity in the *Rhizobium*-legume root nodule symbiosis. *Science*, **185**, 269–71.

BOLWELL, G. P., CALLOW, J. A., CALLOW, M. E. and EVANS, L. V. (1978). Evidence for complementary receptors involved in fertilization in brown seaweeds. *Cell-Cell Recognition*. SEB Symposium No. 32. Cambridge University Press, Cambridge.

BROCK, T. D. (1961). Physiology of the conjugation process in the yeast *Hansenula wingei*. *J. Gen. Microbiol.*, **26**, 487–97.

CRANDALL, M. A. and BROCK, T. D. (1968). Molecular basis of mating in the yeast *Hansenula wingei*. *Bacteriol. Rev.*, **32**, 139–63.

DAY, P. R. (1974). *Genetics of Host-Parasite Interaction*. Freeman and Co., San Francisco.

DUNTZE, W., MACKAY, V. and MANNEY, T. R. (1970). *Saccharomyces cerevisiae:* a diffusible sex factor. *Science*, **168**, 1472–3.

FLOR, H. H. (1971). Current status of the gene-for-gene concept. *Ann. Rev. Phytopath.*, **9**, 275–96.

JOHNSON, R. (1976). Genetics of host-parasite interactions. In: *Specificity in Plant Diseases*, eds. R. K. S. Wood and A. Graniti. Plenum Publ. Corp., New York, 45–64.

HARTWELL, L. H. (1973). Synchronization of haploid yeast cell cycles, a prelude to conjugation. *Exptl. Cell Res.*, **76**, 111–17.

HESLOP-HARRISON, J. (1975a). The physiology of the pollen grain surface. *Proc. Roy. Soc. B.*, **190**, 275–99.

HESLOP-HARRISON, J. (1975b). Incompatibility and the pollen-stigma interaction. *Ann. Rev. Plant Physiol.*, **26**, 403–25.

KNOX, R. B., WILLING, R. R. and PRIOR, L. D. (1973). Interspecific hybridization in poplars using recognition pollen. *Silvae Genet.*, **21**, 65–9.

LEWIS, D., BURRAGE, S. and WALLS, D. (1967). Immunological reactions of single pollen grains; electrophoresis and enzymology of pollen protein exudates. *J. Exptl. Bot.*, **18**, 371–8.

LINDSAY, D. W., YEOMAN, M. M. and BROWN, R. (1975). An analysis of the development of the graft union in *Lycopersicum esculentum*. *Ann. Bot.*, 639–46.

MIRELMAN, D., GALUN, E., SHARON, N. and LOTAN, R. (1975). Inhibition of fungal growth by wheat germ agglutinin. *Nature*, **256**, 414–16.

NASRALLAH, M. E. and WALLACE, D. H. (1967). Immunochemical detection of antigens in self-incompatibility genotypes of cabbage. *Nature*, **213**, 700–1.

NETTANCOURT, D. DE, DEVREUX, M., BOZZINI, A., CRESTI, M., PACINI, E. and SARFATTI, G. (1973). Ultrastructural aspects of the self-incompatibility mechanism in *Lycopersicum peruvianum* Mill. *J. Cell Sci.*, **12**, 403–20.

ROTH, S. (1973). A molecular model for cell interactions. *Quart. Rev. Biol.*, **48**, 54–63.

SINGER, S. J. and NICOLSON, G. L. (1972). The fluid mosaic model of the structure of cell membranes. *Science,* **175**, 720–31.

STEGWEE, D. (1976). Physiology of sexual reproduction in mucorales. In: *Perspectives in Experimental Biology*, ed. N. Sunderland. Pergamon Press, Oxford.

STROI   ce of the toxin-
    k   s. *Proc. Nat. Acad.*

WIESE   n *Chlamydomonas.*

YANA   ., SAKURAI, A. and
                                                                         in *Saccharomyces*
                                                                         y their respective

YEOM   ition of the graft
                                                                         i5–76.